Parallel Processing, 1980 to 2020

Synthesis Lectures on Computer Architecture

Editor
Natalie Enright Jerger, University of Toronto
Editor Emerita
Margaret Martonosi, Princeton University

Founding Editor Emeritus
Mark D. Hill, University of Wisconsin, Madison

Synthesis Lectures on Computer Architecture publishes 50- to 100-page books on topics pertaining to the science and art of designing, analyzing, selecting, and interconnecting hardware components to create computers that meet functional, performance, and cost goals. The scope will largely follow the purview of premier computer architecture conferences, such as ISCA, HPCA, MICRO, and ASPLOS.

Parallel Processing, 1980 to 2020
Robert Kuhn and David Padua
2021

Data Orchestration in Deep Learning Accelerators
Tushar Krishna, Hyoukjun Kwon, Angshuman Parashar, Michael Pellauer, and Ananda Samajdar
2020

Efficient Processing of Deep Neural Networks
Vivienne Sze, Yu-Hsin Chen, Tien-Ju Yang, and Joel S. Emer
2020

Quantum Computer Systems: Research for Noisy Intermediate-Scale Quantum Computers
Yongshan Ding and Frederic T. Chong
2020

A Primer on Memory Consistency and Cache Coherence, Second Edition
Vijay Nagarajan, Daniel J. Sorin, Mark D. Hill, and David A. Wood
2020

Parallel Processing, 1980 to 2020
Robert Kuhn and David Padua

ISBN: 978-3-031-00640-1 print
ISBN: 978-3-031-01768-1 ebook
ISBN: 978-3-031-00065-2 hardcover

DOI 10.1007/978-3-031-01768-1
A Publication in the Springer series
SYNTHESIS LECTURES ON ADVANCES IN AUTOMOTIVE TECHNOLOGY

SYNTHESIS LECTURES ON COMPUTER ARCHITECTURE
Lecture #54

Series Editor: Natalie Enright Jerger, University of Toronto

Series ISSN   1935-3235   Print   1935-3243 Electronic

Parallel Processing, 1980 to 2020

Robert Kuhn
Retired (formerly Intel Corporation)
David Padua
University of Illinois at Urbana-Champaign

SYNTHESIS LECTURES ON COMPUTER ARCHITECTURE #54

ABSTRACT

This historical survey of parallel processing from 1980 to 2020 is a follow-up to the authors' 1981 *Tutorial on Parallel Processing*, which covered the state of the art in hardware, programming languages, and applications. Here, we cover the evolution of the field since 1980 in: *parallel computers*, ranging from the Cyber 205 to clusters now approaching an exaflop, to multicore microprocessors, and Graphic Processing Units (GPUs) in commodity personal devices; *parallel programming notations* such as OpenMP, MPI message passing, and CUDA streaming notation; and seven *parallel applications*, such as finite element analysis and computer vision. Some things that looked like they would be major trends in 1981, such as big Single Instruction Multiple Data arrays disappeared for some time but have been revived recently in deep neural network processors. There are now major trends that did not exist in 1980, such as GPUs, distributed memory machines, and parallel processing in nearly every commodity device.

This book is intended for those that already have some knowledge of parallel processing today and want to learn about the history of the three areas. In *parallel hardware*, every major parallel architecture type from 1980 has scaled-up in performance and scaled-out into commodity microprocessors and GPUs, so that every personal and embedded device is a parallel processor. There has been a confluence of parallel architecture types into hybrid parallel systems. Much of the impetus for change has been Moore's Law, but as clock speed increases have stopped and feature size decreases have slowed down, there has been increased demand on parallel processing to continue performance gains. In *programming notations and compilers*, we observe that the roots of today's programming notations existed before 1980. And that, through a great deal of research, the most widely used programming notations today, although the result of much broadening of these roots, remain close to target system architectures allowing the programmer to almost explicitly use the target's parallelism to the best of their ability. The parallel versions of applications directly or indirectly impact nearly everyone, computer expert or not, and parallelism has brought about major breakthroughs in numerous application areas. Seven parallel applications are studied in this book.

KEYWORDS

history of computing, parallel applications, parallel computers, supercomputers, parallel computing models, parallel programming languages, parallel programming paradigms, parallel programming models, parallel runtime environments

Contents

Foreword by David Kuck

This book covers the past 40 years of parallel computing. To appreciate that statement, consider the 40 years of computing *previous to that*. In 1940, computers barely existed, and they were regarded as special-purpose devices for many following years; not until the middle of that 40-year period did they become commonplace in business and technical areas. Also, in the 1960s, parallel computing hardware began to be developed for special purposes.

Today, serious computing (business and technical), communication (serious and frivolous), and entertainment (now dependent on serious computing and merged with communication) are all rolled into devices that contain multicore parallelism. Before 1940, these were strictly distinct areas, and things remained so for all but a few cases until the period discussed in this book. Of course, today's cell phone or tablet users are unaware of the remarkable fusion of ideas, technical innovations, and integration steps that were required to produce the devices now carried by 2/3 of the world's population. Few people even considered the possibilities before 1980, and those who did would be surprised by what is commonplace today.

Parallelism is not used in all applications, but key areas in technology, business, and entertainment now depend on it for performance. This book covers the hardware/architecture, software, and applications that have been developed to make this possible. Remarkably, special purpose architectures and software methods have not been required to make most of this happen. Although, in many cases, specialized hardware feeds market growth by delivering the best performance increases. Also, some application domains have grown by using embedded components, e.g., parallelism supports computer vision functionalities built into medical devices, robots, vehicles, security systems, etc. While each application area indeed exploits distinct hardware and software configurations, a great deal of shared or similar underlying technology is used.

In high-performance computing, the dependence on parallel technology has become widespread, and in many cases absolute. Computing has become the "third branch" of science, joining theoretical and experimental science as an equal partner. Scientists now use very accurate parallel simulations to understand how the world works and explore how it can be changed in various ways. Multiple Nobel prizes have depended on computational results. At the other pole, many movies, most TV commercials, and a world of games totally depend on parallel computation for their existence, as do all sorts of social networking applications.

By blending computing, communication, and entertainment in the past 40 years, technologists have created a world that seems quite different from the one that existed before 1940, and most of the change has occurred in the period covered by this book. Many of the innovations necessary to support this new world are discussed in some detail herein. This book also probes the questions of how these things came to happen. The changes depend on many factors beyond technology, but it is important to understand the interplay of technical ideas with the business world.

Necessity is no longer the "mother of invention." That idea has been expanded to include the possibility of doing something that seems useful or is even simply marketable. Fueled by a host of hardware and software building blocks, software people can create applications that sweep disruptively into the marketplace, even though they weren't regarded as necessities or thought to be of much value until they appeared. This same mentality dominated computer architecture in the 1980s and is reported in detail in this book. Software has mostly followed the demands of hardware, but domain-specific software tools and programming notations have been created to help specific application areas.

The dynamics of the applications, architecture, and software triangle are constantly changing, driven by many factors. Many sources of business capital growth are now present beyond the traditional sale of equity and profits from end-user sales, ranging from venture capital pushes to internet advertising revenue. This has sometimes driven product development in new ways. Similarly, the fact that hardware has reached physical limitations (e.g., signal propagation times) reduced the rate of computer speed growth. This increased the demands on system architecture and software parallelism to maintain system speed growth. Parallelism and multiprocessing in Systems-on-a-Chip (SoCs) are much intertwined with speed, power, and thermal effects, so they sometimes provide system-design options for the market side necessities of portability and easy use that have helped drive the field.

This book provides a nice balance of discussions covering the technologies of hardware/architecture, software notations, and applications areas, framing many advances in parallel computing. Just as the sources of investment money have changed, so have the interactions among these three technologies. Thus, it is sometimes hard to understand or explain just why the field moved one way or another, but an understanding of the technologies discussed here is a prerequisite, because multiple factors are usually involved. Due to the rate of technical change, some product changes move so quickly that technologists are surprised, and investors are shocked.

Where the field will be 40 years hence is impossible to predict because 40 years ago few specialists could have imagined where we would be today, and given the much more complex technology today, it seems even less likely that we can make reasonable predictions now. But looking back is always a good way to judge progress and the current state of affairs. By that method, we can expect a continuing growth of parallel computing in many surprising and innovative directions. Based on the coordinated growth of these three technologies, more areas will yield to one kind of parallel computation or another, so whatever hardware technologies survive, performance and application utility will continue to grow without bound.

I was personally impressed by the fact that even though I worked through this entire period as a parallel computing advocate, the authors have revealed and explained interesting relationships that I had forgotten or never recognized myself. I believe that students new to the field, experienced practitioners, and historians of technology will all benefit from this book. As with any historical review, one cannot really understand how the world moves forward without considering several insightful historical perspectives. This book is a serious contribution to those perspectives. Finally, since both authors were my students in the 1970s, it is especially rewarding to be learning from them about a field in which we have all been active for so long.

David Kuck
Intel Corporation

Preface

The last few decades saw parallel computing come of age. We were fortunate to be throughout our careers in close proximity to these developments, among the most exciting in the history of technology. The numerous challenges remaining should not make us forget the immense successes that enabled, in only 40 years, the transition of the field from a small research area to the norm of today and tomorrow.

Clearly, now is an excellent time to look back and summarize the evolution of parallel computing. This and our realization that 40 years had passed since we gave a tutorial at the International Conference on Parallel Processing (ICPP) motivated us to undertake this short history. The tutorial is our starting point and the basis for our focus on hardware, programming notations, and applications. Initially, we intended this booklet to be even shorter than it became, but we did not fully realize until we were in the middle of the work how long a time is 40 years and how incredibly productive those working in this area have been.

We assume that the reader is familiar with parallel computing technology, both hardware and software. We tried to use widely accepted terminology and, in most cases, we do not give definitions. And we do not include references to the literature for some of the machines discussed; often only marketing driven documents and user guides remain.

However, most concepts are not advanced and the background given by any introductory parallel computing course should suffice to follow the presentation and for most machines discussed, internet search engines should be able to identify good descriptions.

Readers are encouraged to read in the order that their curiosity leads them with the note that Chapter 2, on parallel hardware, follows the evolution of parallel hardware through time.

We would like to express our gratitude to our common Ph.D. thesis adviser, David Kuck, for agreeing to write the foreword and especially for recommending us to Tse-yun Feng, the organizer of ICPP. We also would like to thank Stott Parker for his invaluable recommendations and the people from Morgan & Claypool publishers, especially Mike Morgan, Christine Kiilerich, and Natalie Enright Jerger for their help and encouragement during the preparation of the manuscript.

Robert Kuhn and David Padua
September 2020

Acknowledgments

The authors wish to thank those that read and gave comments on the drafts, and steered us to other information important to share in passing on the history of parallel processing. In particular, we thank Stott Parker and David Kuck. We thank those that have helped us fill key gaps, such as Michael Booth, Geoffrey Fox, Mark Furtney, and Bruce Leasure. Their personal recollections gave us key perceptions from their different points of view. The authors wish to thank all of the experts in parallel processing we have worked with over the last 40 years. For example, and a small part of a long list, from the early days of parallel applications: Rosario Caltabiano, Nick Camp, Gilles Deghillage, Roberto Gomperts, Marty Itzkowitz, Martin Lewitt, Larsgunnar Nilsson, Yih-Yih Lin, and Madan Venugopal. They and many others have all contributed to progressing the field from its nascent state in 1980 to the large and growing technology it is today. But they also have contributed time from their careers and families to make parallel processing a deeply rewarding personal experience for us. We want to most strongly thank those we have contacted recently to help guide this book.

CHAPTER 1

Introduction

In 1981, the authors edited the *Tutorial on Parallel Processing* containing papers on hardware, programming languages, and applications [KuPa81]. It was one of the first collections of papers on parallel processing and tried to capture the state-of-the-art. This book investigates what has happened since then. To characterize parallelism in this period there are five classes of developments that recur.

1. Performance improvements: Speedup is THE measure of parallelism. One enabler of speedup is the hardware performance gains of more than 3 orders of magnitude over the last 40 years. Step-by-step, how did that occur? Item 5 below is the other enabler required for success: the development of applications that speedup with the hardware.

2. Popularization: Parallelism would remain useful to the R&D sector alone had it not been that the hardware in everyday use, commodity hardware, has also become parallel. The most widely used parallel commodity hardware are general purpose units (GPUs) and multicore CPUs. Between these two, GPU applications consistently use parallelism. For multicore microprocessors, although more slowly than we would like, applications are becoming parallel (see Appendix A).

3. Incremental consistent innovation: We have seen many innovations in parallelism in 40 years. These good ideas have continued forward as building blocks of the next-generation systems, often in hybrid forms.

4. Parallel programming notations broaden and diffuse: We show that the main ideas behind current parallel programming notations existed in 1980. But, each one needed, first, to be broadened to be easily applicable in many applications. Second, for developers to learn and apply parallel programming notations takes time, years in fact. That is, parallelism takes time to diffuse into applications.

5. Growth in the number and performance of scalable applications: Parallel applications now impact nearly everyone every day. Reaching this level of impact takes years of version after version improvements. In addition to parallel hardware with good performance being available there must be motivations from the users' side to apply parallelism. The four frequent parallel application growth drivers we have found are:

growth of data, adding new layers of scalable parallelism, new algorithmic methods, and time-to-solution requirements.

These five classes of development have resulted in the broad adoption of parallel processing we see today. (Over these years parallel processing has also had its share of myths and misconceptions. Some of these are captured in Appendix A.)

The goal of identifying these developments is to help us better understand the problems we face today. One does need not need to search far to find numerous challenges for parallel processing.

- On the application side, how to use new parallelism to improve current applications and attack new applications? Someone passionate about parallelism might say, how do we defeat Amdahl's Law every time?

- On the hardware side, over time Moore's Law has solved multiple problems without parallelism. Many say Moore's Law is ending. Can parallelism step up to enable computing to grow through this challenge?

- On the programming notation side, invention, broadening, and diffusion have worked, but they have worked slowly. Are we still on a steady, efficient growth track?

Today, especially, parallel processing is a broad term. For purposes of historical analysis we used a wide definition to note interesting developments across the field. At the same time, to apply a limit to what should be included, by parallel processing we mean multiple hardware units working simultaneously to solve a particular problem faster. Note that this goal may overlap with the use of other technologies in some applications described below. For example, parallel processing often coexists with distributed processing and multiprogramming.

The rest of this chapter describes the organization of the book and presents a quick overview of the evolution of hardware, programming notations, and applications during the last four decades.

1.1   OVERVIEW OF THE BOOK

In the three subsequent chapters, this book covers the evolution from 1980 to our times of the binding of parallel applications, to parallel hardware, through parallel programming notations. Chapter 5 overviews where are we now and asks where can we go in the short term.

Chapter 2 shows the evolution of parallel hardware in several dimensions. In 1980, the Cray-1 had a single processor with a peak performance of 160 megaflops. Today at the high-end, the number of processors per machine has grown by a factor of hundreds of thousands. As a result, the Summit cluster (DOE/IBM) is approximately 188,000 teraflops. (See [TOP500] for the fastest performance systems today.) Over four decades, the high-end pipelined vector processors, Single Instruction Multiple Data (SIMD) machines, and shared memory machines of the 1980s, along with the companies that did the pioneering work, with a few exceptions, have disappeared. Now, even at the low-end vector pipelining, SIMD, and shared memory model (SMM) designs are present in commodity microprocessors, GPUs, and "AI chips" referred to here as Deep Neural Net Processors (DNNPs). Figure 1.1 shows a few of the important parallel processors that have appeared over this period. Although not initially parallel, the PC introduced by IBM in 1981 has been an important pathway for the expansion of parallelism. The original PC with its 4.77 MHz Intel 8088 processor evolved into today's laptop, desktop, and smartphone systems, which almost universally contain multicore processors and GPUs operating at around 4 GHz. Although the price for the largest systems is still very high today, the low price per megaflop provides more than the power of a Cray-1 in every smartphone.

Chapter 3 shows that the seeds of parallel programming sown by 1981 have become broader and have diffused to common usage. The mutual exclusion constructs including the monitors of Concurrent Pascal, the message passing primitives of CSP, the DOALL construct of the Burroughs FMP, and the simple fork/join operation of 1963 have evolved into the large number of powerful primitives in a handful of standard programming notations today: MPI, OpenMP, TBB, standardized threading primitives, CUDA, OpenCL, and libraries and flowgraphs encapsulating parallelism. On the other hand, the focus on getting the highest performance, which characterizes parallelism, has blocked progress in some programming notations. High-level programming notations require powerful compilers for efficient implementation, but today's compiler techniques still cannot match manual coding with low-level programming notations, such as MPI, which has been called the "assembly language of parallel programming." Note that, although low-level, MPI has greatly added to the standardization of the message passing notation.

Chapter 4 shows that the abundant availability of parallel hardware has meant that parallelism has come into broad use so that nearly every person has benefitted from parallel processing in one application or another. Many of these applications are datacenter applications used in a wide variety of industries. For these, often the benefit to end-users is indirect through engineers, scientists, and designers applying the results produced by parallel applications. There are also many applications that directly benefit end-users running on the laptop, desktop, handheld, and embedded hardware that has emerged since 2000. Success has not been rapid; most applications have made parallel processing practical step-by-step over several years.

Seven parallel application areas are discussed in this book. In each of the application areas, a striking breakthrough due to parallelism that impacts millions of end-users has been achieved. The columns of Table 1.1 correspond to problem domains. Most of these applications have implementations for distributed memory, shared memory, and GPU architectures. The table illustrates that over time more parallel applications have been deployed to end-user devices. Of course some applications will stay in the datacenter because they need massive parallelism or they produce results that must be interpreted by domain experts. Large parallel computing clusters are usually found in laboratories for use by researchers; seismic analysis is an exception. Medium clusters are usually found in industry R&D datacenters. Applications that can be deployed to end-user devices, such as vision and speech applications, are often developed in datacenters and may even be deployed there for business strategy reasons.

There are four appendices. Appendix A covers some of the myths and misconceptions about parallel processing. Appendix B contains bibliographic notes. Appendix C is taxonomic notes. Finally, Appendix D outlines the 1981 tutorial on parallel processing.

Chapter 5 asks if recent parallelism reflects historical parallel processors. For example, do the features of deep neural net processors inherit features of prior parallel processors? Are we making progress toward exascale clusters? What is happening with the biggest multicores today? Some architectural concepts in GPUs in 2020 go back to 1980, but with a very different implementation. A subject central to parallelism's progress asks, if Moore's Law is slowing, where can fabrication and packaging technology go. The last section asks whether parallel system architecture and software can continue to scale-up using fabrication and packaging in the near term.

Table 1.1: Some parallel applications over time on clusters and commodity hardware

Date	1975	1985	1995	2005	2010	2010	2015	Primary Active Sector
Large clusters	Seismic processing *3D Survey (1971)*			Genomics *Sequence human genome (2001)*	Molecular dynamics *Protein folding (2010)*			Research laboratories
Medium clusters		Dynamic FEA *Full car crash simulation (1983)*	Photorealistic rendering *Full length animation (1995)*					Industrial R&D laboratories
End user and Embedded devices			Realtime rendering *PlayStation (1995)*			Image processing		Consumer products
						Image recognition *Autonomous vehicle (2010)*	Speech recognition *Word error rate <5% (2015)*	

(Breakthrough results of applications in Chapter 4 are shown by year in columns with the application method in bold and the result in italics. Rows indicate typical parallel processors used today for each application.)

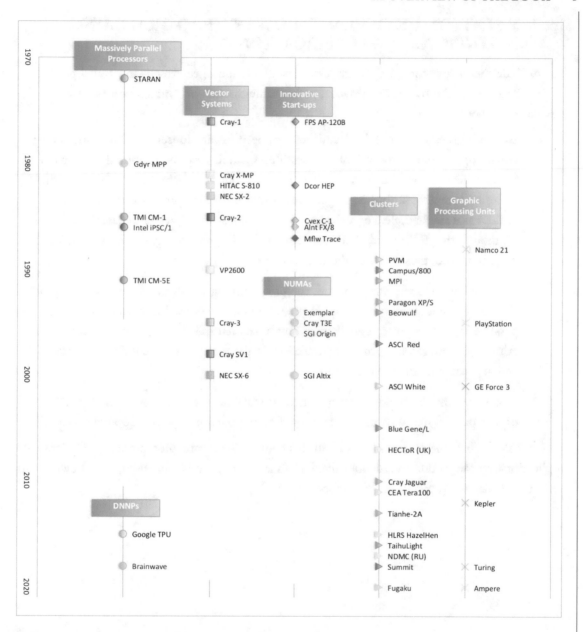

Figure 1.1: Some parallel system breakthroughs.

1.2   RELATION BETWEEN HARDWARE, PROGRAMMING NOTATIONS, AND APPLICATIONS

To conclude the introduction, important lessons can be found in the trends in parallel processing. Table 1.2 illustrates the trends in hardware, programming notations, and applications that are described in the book.

- At the high end, the trend in hardware has been toward looser coupling, which increases opportunities for scalable applications. Coupling can be measured in terms of the number of processor clock cycles to communicate data between processors or the cycles to complete a pipeline stage. For a vector or SIMD processor the coupling is as little as a single cycle. For a cluster, it can be thousands of cycles to pass a message. (Although the SIMD and VLIW systems shown in grey are tightly coupled and reverse the trend, these reverses did not survive.)

- In the table, for each hardware row, the programming notation most appropriate to that hardware is shown. The trend in programming notations has been from implicit parallelism, such as autoparallelization, toward more explicit programming notations. Although more recently, implicit parallelism, in the form of macrodataflow and application specific libraries, has been used.

- In applications, the trend has been to increase scalability over time by taking advantage of new parallel hardware and better programming notations as they became available.

Table 1.3 lists the major terms used in this book for hardware, programming notations, and applications, by the section that defines them. If an acronym for a term is used, it is shown in the table. (Refer also to Appendix C for taxonomic notes.)

Table 1.2: Major classes of systems with programming notations and applications dominant on them (grey rows indicate classes that are no longer available or little used)

Period	Hardware	Coupling	Principal Parallel Programming Notation	Implicit to Explicit	Example Application(s)	Low to High Scalability
<1980	Early SIMD	< 5 cycles	Array syntax	Explicit	Image processing	High - Array of pixels
<1980	Cray vector processors	1 cycle, e.g., mult-add dot product	CFT vectorization BLAS libraries	Implicit (plus directives)	FEA Seismic	Medium - Vectors of elements High - Parallel shots
1980–1985	Cray XMP vector-multiprocessor	~10 cycles - Memory write, Semaphore register set, Memory read	Cray Microtasking	Explicit	Computer graphics Molecular dynamics	Medium - Vectors of polygons Vectors of atoms
1985–1990	VLIW	1 cycle - Chained function units	Trace Compiler	Implicit	Automatic speech recognition	Low - Finite state transducers
1985–1990	MPP SIMD	< 5 cycles - Neighbor transfer	Array syntax	Explicit	Image processing	High - Array of pixels
1990–1995	NUMA	~100 cycles - Flush cache line Load cache line	MPI, OpenMP	Explicit	Gene matching Drug discovery	High - Partition gene DB High - Partition Molecule DB
1990–1995	DMM	> 10K cycles - Message pass	MPI	Explicit	Automatic speech recognition	Medium - Deep learning & multitasking neural net recognizers
2000 – Today	Hybrid Clusters of Multicores & PUs	Mixed loose to tight - SIMT	MPI, OpenMP, OpenACC, CUDA	Explicit	Of those above - Seismic, Comp. Chem., Bioinformatics, ASR, Computer vision	High - As above
2010 – Today	Deep Neural Net Processors	Tight < 5 cycles	Macrodataflow, parallel computing libraries	Explicit/Implicit (Macrodataflow is implicit)	CV, ASR, increasing in Bioinformatics and CC	High scalability

Table 1.3: Table of key parallel processing terms used in this book (the section in which the term is defined or first used is given in the reference column)

Hardware Terms, Chapter 2			Hardware Terms, Chapter 3			Application Terms, Chapter 4		
Term	Acronym	Reference	Term	Acronym	Reference	Term	Acronym	Reference
Classic parallel processors	--	2	*Array notation*	--	3.1	*Seismic data processing*	--	4.1
Single Instruction Multiple Data	SIMD	2.1, 2.4.2	Triplet notation	--	3.1	Seismic migration	--	4.1
Pipelined vector processor	--	2.1, 2.3	For all	--	3.1	*Finite Element Analysis*	FEA	4.2
Shared Memory Model	SMM	2.1	*Automatic vectorization*	--	3.2	Implicit FEA method	--	4.2
Innovative start-ups		2.4	Autovectorizing compilers	--	3.2	Explicit FEA method	--	4.2
Dataflow	--	2.4.1	Autoparallelizing compilers	--	3.2	*Computer-Generated Imagery*	CGI	4.3
Non-Uniform Memory Access	NUMA	2.4.1	Dependences	--	3.2	Polygon rendering	--	4.3
Very Long Instruction Word	VLIW	2.4.1	Directives	--	3.2	Photorealistic rendering	--	4.3
Massively Parallel Processor	MPP	2.4.2	*Single-Program Multiple Data Programming*	SPMD	3.3	*Bioinformatics*	Bio-info	4.4
Interconnection networks	--	2.5.4	Branch divergence	--	3.3	Gene sequencing	--	4.4
Modern hardware			*Message passing primitives*	--	3.4	Gene matching	--	4.4

Distributed Memory Model	DMM	2.5.1	One-sided message passing	--	3.4	Computational Chemistry	CC	4.5
Multicore	--	2.5.5, 5.4	Partitioned Global Address Space	PGAS	3.4	Molecular Dynamics	MD	4.5
Manycore	--	3.5	Shared memory primitives	--	3.5	Ab initio methods	--	4.5
Simultaneous Multi-threading	SMT	2.4.3	Thread reuse	--	3.5	Drug discovery	--	4.5
Graphic Processing Unit	GPU	2.5.5, 5.5	Do all	--	3.5	Computer Vision	CV	4.6
Field Programmable Logic Array	FPGA	2.5.5	Template library	--	3.5	Image processing	--	4.6
Application Specific Integrated Circuit	ASIC	2.5.5	Dataflow	--	3.6	Image recognition	--	4.6
Commercial Off The Shelf	COTS	2.5.1	Macrodataflow	--	3.6	Image understanding	--	4.6
Single Instruction Multiple Threads	SIMT	5.5	Parallel computing libraries	--	3.7	Automatic Speech Recognition	ASR	4.7
Deep Neural Net Processor	DNNP	5.2	Application specific libraries	--	3.7	Probabilistic methods	--	4.7
Thermal Design Power	TDP	5.7				Deep neural nets	DNN	4.7

CHAPTER 2

Parallel Hardware

Seymour Cray was quoted as saying "If you were plowing a field, which would you rather use? Two strong oxen or 1,024 chickens?" [WikCray]. This quote vividly illustrates the evolution of parallel processing from 1980–2020. Figure 2.1 breaks down the timeline into three overlapping waves of parallel systems.

The blue bubble is the first wave, vector supercomputers. Inside the bubble, the blue line shows that Cray supercomputer performance increased by over a factor of a thousand in the 15 years from the Cray-1 to the Cray C90. By the early 1990s, turning point numbered 1 in the figure, pipelined vector supercomputer peak performance was below the peak performance of large Distributed Memory Model (DMM) clusters showing Cray may have underestimated the power of scalability. He was not wrong at that time. Vector supercomputers were able to reach a higher percentage of peak performance and had more applications running at high speed, but the future was not looking good for vector supercomputers. (Cray had a way of making figurative quotes at the right moment. For example he said, "Parity is for farmers," when challenged by memory error rates compared to machines that had error correcting memory.) Around 1988, the Cray Y-MP/832 was the last Cray vector processor to hold the number one position in the TOP500. Followed by the Fujitsu VP2600/10 and the NEC SX-3-44, they were the last vector supercomputers with modest multiprocessor configurations, i.e., less than 32 processors. In 1993, the TMC CM-5 with 1,024 processors became the first non-vector system to reach the number one position. All systems after had hundreds to thousands of processors. (The bottom point of the vertical descending dotted line, the "tail," on the performance lines in the figure show the performance of a single processor in the system.) A single C90 processor in Emitter-Coupled Logic (ECL) was only twice as fast as a CMOS GPU. This is hardly price-performance competitive.

Meanwhile, the second wave, machines produced by innovative start-ups, are shown as a green bubble in Figure 2.1. Mini-supercomputer manufacturers were at first successful start-ups. Convex performance improvement over time is representative of mini-supercomputers. (It is shown in the dark green line.) Their performance growth over time looked impressive. However, other designs soon overtook mini-supercomputers. Consider the 1991 end of the line mark that shows the performance of an eight-processor system connected with a crossbar switch. The bottom of the vertical dotted line descending from the end of line mark shows that the performance of a single processor is below the performance of a single GPU. Because the cost-performance of the "killer microprocessor" was apparent by that time, and given the time and resources needed to design the next generation of a custom architecture, wave two companies, such as Convex, switched to NUMA microprocessor-based architectures or went out of business. For example, Convex worked with HP to produce the Exemplar NUMA architecture. NUMA supercomputer performance improvement over time kept pace with clusters for over 10 years, as represented by SGI NUMA systems, the light green line in Figure 2.1. The SGI Origin 2000 evolved through several models of NUMAlink systems into the Altix 4700. At the beginning of this period the performance gap to a cluster could be rationalized because NUMA systems are easier to program than a DMM cluster. But by the end of this period, turning point numbered 2 in the figure, around 2010, the higher performance of clusters became more appreciated because message passing technology had diffused into more high-end applications. The Altix 4700 was the last of the SGI NUMA breed.

The third wave, the red bubble in Figure 2.1, consists of two lobes. The upper lobe contains clusters taking over as the highest performing supercomputers. One can see the performance trend of clusters in the dark red line at the top of the red bubble. As just mentioned, in 1993 the CM-5 reached the number one ranking in the TOP500. Performance improvement of clusters increased rapidly. Around 2008, the first GPU accelerated cluster, Roadrunner, reached the number one ranking and the "accelerated cluster" performance trend takes a slightly steeper slope at that point. Clusters are expected to reach exaflop performance by the early 2020s.

The lower lobe of the third wave resulted in the popularization of parallelism in commodity consumer systems that became unstoppable around 2006 with the introduction of multicore microprocessors and manycore GPUs. From turning point 1, around 1992, one notes the steep rise in performance of both microprocessors and GPUs. This is represented by an orange line for Intel microprocessors and a red line from the Namco System 21 of 1988 to the Nvidia Turing system of 2018. There was not much parallelism in these processors early on but they were commodity items. For example, the Sony PlayStation was introduced in 1994–1995, which ultimately sold over 102 million systems, contained an SGI MIPS R3000 microprocessor in parallel with a Sony GPU. In addition to high-speed game consoles, some professional workstations contained multiprocessors. But, the big turning point was in 2006 when the dual core microprocessor was launched. From then on, multicore and GPUs were used in every laptop, and then in many embedded systems. By 2020 the top multicore processors contain around 64 cores and GPUs contain over 4000 small cores. (The performance gains of microprocessor technology are discussed in more detail in Section 2.2.) Around 2015 DNNPs for processing neural nets started to appear. It is hard to put them on the same gigaflops scale with the systems above because they compute with shorter representations (32-bit floating point down to 8-bit integer arithmetic) than those typically used by scientific-engineering applications, but for their special purpose they can be over 30 times the performance of a GPU and at wafer scale integration, which one DNNP start-up (Cerebras) is already using, may contain over 100K small cores.

Behind these improvements in performance, in addition to parallelism improvements such as more processors and more parallelism inside the processor, two main trends characterize the

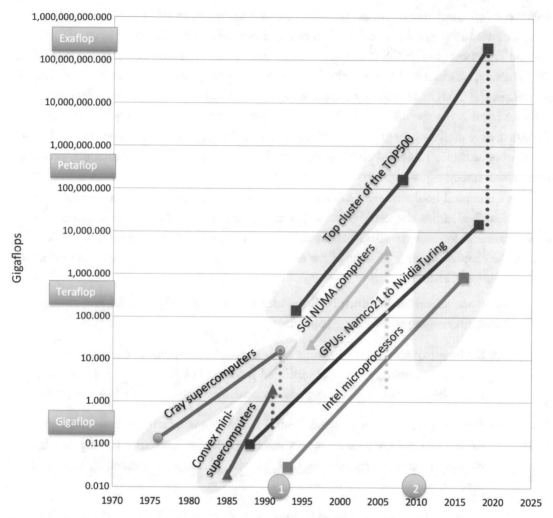

Figure 2.1: Peak performance of major parallel architectures over time. (The blue bubble is vector supercomputers. Cray supercomputers, the blue line, is representive. The green bubble is mini-super-compters. Two lines are representative: Convex computers, darker green, and SGI NUMA computers, lighter green. The red bubble is current parallel systems. There are three representative lines: the top cluster of the TOP500, dark red; Intel microprocecssors, orange; and GPUs from the Namco System 21 to the Nvidia Turing. Around 2010 GPUs and microprocesssors become parallel and enter the third bubble.)

evolution of all computer hardware, including parallel hardware, in the 1980–2020 period. First, is the increase in integration which in turn changed the hardware technology from BJT (Bipolar Junction Transistor) to CMOS (Complementary metal–oxide–semiconductor). In the case of parallel hardware, we observe that while the pipelined vector processors of the first wave used bipolar technology, the CMOS technology transition started in the second wave. The second computer hardware trend is the evolution toward commodity processors. We can observe that the machines of the first wave used custom processors, during the second wave both custom and commodity processors were used (see Tables 2.1 and 2.2), and, finally, during the third wave commodity processors are used universally.

In a technical book one would like to show clear technical reasons to construct the parallel systems described in this chapter. We found that in addition to the technical causes that are covered below, economic forces have also intervened. To clear the path for the technical discussion, let's preview the chapter in terms of some of the economic reasons that have played a role in changing parallel processing history.

- **Win-win** situations in which both the user and creator of the parallel processor are happy for one or more of these reasons.

    ○ **Savings** in applications: In some cases the application was not feasible without parallel processing. The vector pipelined processors of Section 2.1 made applications like 3D seismic processing and car crash simulation feasible (see Chapter 4).

    ○ **Price or performance** increase: Parallel hardware manufacturers have often gone to new technology because it provides better performance or a lower price. Increasing scale of integration over time, Section 2.2, has been compelling to parallel processing.

    ○ Commitments to the **installed base** of customers. (Doing something to keep a customer is win for the manufacturer and one hopes for the customer. Cray Research, Inc. stayed with vector supercomputers until they went out of business, even though they prudently and successfully developed other parallel processors; Section 2.3).

- **Conviction**: To do something beyond technical or monetary advantage, often because one thinks it will be the right thing in the end.

    ○ **Investing assets**, either one's capital or one's time: Sometimes the saving to society or one nation is so strong that an investment is made to find out if a solution is even feasible. A clear case is David Shaw who saw that molecular dynamics was slow to reach its major goals because enough parallelism was not being applied.

D. E. Shaw provided the capital to fabricate ASICs, Section 2.5.5, and the Anton computer (Section 4.5).

○ **Engineers – Engineer**: A management theory holds that one should hire the best people, new and better solutions will be found. It is often subtle to see examples from the outside because it consists of many small innovations. Beowulf, for example, got engineering clusters right, and cluster computing took off (Section 2.5.1).

• Taking advantage of the **economic climate**.

○ When large **market opportunities** appear, new parallel solutions often appear as well. Mainstream computer companies entered parallelism because of that opportunity (Section 2.4.5). GPUs successfully addressed the video game market. Then, they seized the general-purpose GPU (GPGPU) market opportunity when it appeared before them (Section 2.5.5).

○ **Venture capital** companies are a multiplier of this process. When the opportunity is large and a good business plan is made, they provide the bootstrap funding of start-ups, with significant return on investment anticipated. Considerable venture capital appeared in the second wave (Section 2.4). In hindsight, one would think too much capital for the actual market opportunity.

• **Mistakes** – On the downside …

○ **Technology was flawed**: Either it didn't provide the expected performance, it didn't prove stable, or it was too costly to the final product. SIMD arrays (Section 2.4.2) had the potential for massive speedup, which never materialized.

○ **Product didn't sell**: Business plans have often been wrong. Sometimes it's not clear why and the catchall reason is "it didn't sell."

Economic reasons aside, let's outline this chapter in the technically logical order. We start with the context of the four decades from 1980–2020. Section 2.1 looks back at parallel processing in 1980. Section 2.2 reviews the evolution of the scale of integration and clock rate. Then, the body of the subject, the evolution of parallel hardware in this period is discussed as the three overlapping waves shown in Figure 2.1. Section 2.3 covers pipelined vector processors; Section 2.4 is about innovative start-ups; Section 2.5 covers the transfer of parallel processing of the prior waves into new directions that remain today: clusters of microprocessors and commodity parallel systems for personal and embedded uses. As a summary, Section 2.6 lists some common parallel processing design patterns.

(To learn more, many books about various aspects of parallel processing are available to the reader. In addition, Appendix B has a list of online historical sources that the reader may wish to pursue.)

2.1   PARALLEL PROCESSING BEFORE 1980

By 1980 there were four types of parallel systems. Compared to today's systems they are primitive, but they were major breakthroughs in their day. (The 1981 tutorial had a more detailed view of parallel processing then. See Appendix D for an outline of the tutorial.)

- Pipelined vector processors whose most notable examples by 1980 were the CDC STAR-100 [HiTa72], the CDC 205 [Shoo80], and the Cray-1 [Russ78]. These were the production supercomputers of the era. For example, the CDC STAR-100, of 1974, had a peak performance of 100 megaflops and contained only 32 MB of magnetic core memory with an access time of 1.28 microseconds. In contrast, today's cell phones are in the gigahertz range and have more than two orders of magnitude more memory. One of the things that made the Cray-1 unique from the others is that it introduced vector registers instead of memory-to-memory vector operations. Program-controlled caching of intermediate result vectors was possible. And, this enabled the clock to be made faster because the wire lengths ALU-to-registers became shorter. A faster clock made scalar processing faster. (Amdahl's Law was on everyone's mind at that time.) Over 100 Cray-1s were sold. It was one of the most popular supercomputers ever.

- Single Instruction Multiple Data (SIMD) machines which by 1980 were represented by Goodyear MPP [Batc80] and Burroughs BSP [Stok77]. SIMD systems had shown promising results in R&D organizations for a few important applications. The BSP was the follow-on to the ILLIAC IV and featured "conflict-free memory access" which allowed rows, columns, and diagonals of a 2D array to be accessed in parallel. (To implement conflict-free access the number of memory banks, 17, was relatively prime to the number of PEs, 16.) SIMD designs faded from the scene by the mid 1990s until revived by DNNPs.

- Shared Memory Model (SMM) machines whose 1980 representatives were S-1 [Widd80], Cm* [SwFS77, SBLO77], and Burroughs FMP [LuBa86]. An interesting example was the S-1 Mark IIA which had a 15 MIPS processor and microcoded instructions as complex as an FFT instruction. However, none of these systems were more than prototypes. Now practically all microprocessors are SMM multicores.

- Dataflow. There was much excitement about dataflow machines in 1980 [DeMi75], but the potential never materialized in the form of commercially profitable machines. The tutorial classified the Denelcor HEP [Smit78] as a dataflow machine although it really was not. The HEP was, however, a promising concept and a few systems sold commercially.

There were few other parallel systems in that era, for example the TI-ASC [Wats72] and the ICL DAP [Redd73], but not many. In addition there was considerable momentum toward building special purpose parallel processors.

As a glimpse of what was to come, there were several DMM installations at the time; one type used the IBM channel-to-channel adapter to connect mainframes. Although DMM systems were the subject of some parallel processing research [Ston77], such systems were frequently throughput oriented and not parallel processing solutions [ChLi78].

2.2   CLOCK SPEED AND GROWTH IN SCALE OF INTEGRATION

Supercomputer integrated circuit technology has changed dramatically in the last 40 years. In 1981, the Cray-1 and other top machines were built with ECL small-scale integration chips. Today, CMOS technology is almost universally used and a truly dramatic increase in integration leading to chips with more than 7 billion transistors. This has made it possible for nearly all machine organizations commercialized in 1980 to be reincarnated in current microprocessors: SIMD execution in GPUs and vector extension instructions, SMM multicore processors, and pipelining in instruction processing as well as floating point arithmetic operations.

Since 1980, clock rates have increased several orders of magnitude. The original PCs were around 4 MHz. (The Cray-1 was 80 MHz and was limited by the speed of propagation in copper interconnects.) With each generation Intel would alternately improve the fabrication process and then improve the microarchitecture. This was so regular it resulted in Moore's Law and became known as the Tick-Tock model. Despite this growth, and perhaps the most important event in machine organization since 1980, is that clock cycles per second stopped increasing around the middle of the decade 2000. This signaled the end of Dennard Scaling of CMOS. Dennard Scaling says roughly that if a CMOS circuit's density is doubled, the clock speed can be increased roughly 40% without increasing power consumed. The problem was that as the feature size decreased further the CMOS power consumption started increasing dramatically due to current leakage.

Figure 2.2 illustrates the growing performance of the Intel series of processors from 1993–2019. Although the clock frequency stalled around 2004, vector instructions and multicore has kept the peak gigaflops increasing at the same rate. For example, in 2017 peak performance grew faster than any year before thanks to the production of the largest Xeon processor that, at the time, had

28 cores implementing the 512-bit advanced vector extensions (AVX-512). This illustrates that parallelism could allow performance to grow without exceeding chip power limits. (Although the quantity shipped of this particular processor was probably low compared to the other mainstream Intel Xeon servers and the mainstream Intel Core series desktop and mobile processors, growth of cores in microprocessors continues today; see Section 5.4.) Some claim that today we are close to the limits of parallelism scaling; others argue that is not so (see Section 5.7). More claim that we are close to the end of Moore's Law (see Section 5.6) citing that decreasing the lithography has become more difficult for Intel [Cutr18]. The Tick-Tock model has been slowed to three phases: Process, Architecture, and Optimization.

In either case, it has been noted often that increasing the number of cores or the vector length is not as effective for general computing as increasing the clock rate. The mega-challenge today for the parallel processing field, not just hardware manufacturers, is to use parallelism to continue to overcome stalled clock rates and slower decreases in lithography, and to make parallelism effective for almost any application.

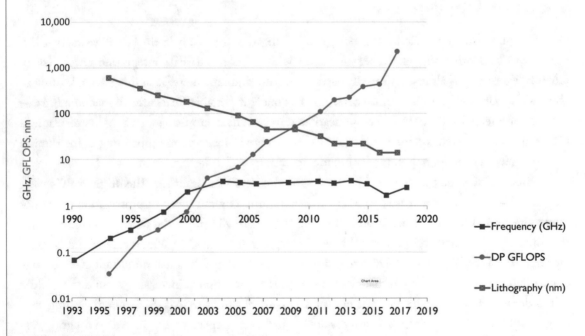

Figure 2.2: Intel microprocessor progression from 1993–2019 (updated from [ChFP08]). (The lithography feature size (in nm) has decreased steadily. The peak double precision gigaflops have increased steadily despite the peak in clock frequency (in GHz) at around 3 GHz.)

2.3   VECTOR SUPERCOMPUTERS

The first wave, from the late 1970s to the late 1990s, was pipelined vector processors. This could be called the age of big iron supercomputing because there were few small computer companies that could afford to design, build, and support these big systems. Cray Research, Inc. was an exception and they often had to convince the large enterprises that bought a Cray that they should spend several million dollars with a relatively small company. (By the third wave, small companies were more acceptable. A small integrator of COTS (Commercial Off-The-Shelf) hardware and open source software can build a cluster system.) In the U.S., Cray dominated as the TI-ASC, CDC STAR-100, and the CDC Cyber 205 were withdrawn from production relatively quickly. Then, CDC spun off the ETA Systems Company, whose ETA-10 was produced from 1986–1990 and then it was withdrawn also.

In Japan, three companies entered the market in 1982–1983, later than Cray in this wave: Fujitsu, e.g., the FACOM VP; Hitachi, e.g., the HITAC S-810; and NEC, the SX Series. They successfully transitioned from first wave vector supercomputer architectures to massively parallel microprocessor architectures. Their systems successfully reached the top of the TOP500 and extended this wave of processors to the recent past with systems such as the NEC SX-Aurora as stated below, the Hitachi pseudo vector processor in 1992 also below, and recently Fujitsu's ARM A64FX microprocessor with Scalable Vector Extensions (SVE) (Section 5.4).

The main period of vector system dominance had a large impact on parallel programming notations and applications, as described in Chapters 3 and 4. Autovectorization, autoparallelization, and shared memory programming notations were employed for the first time on big iron systems. Many successful parallel applications in industries such as national research labs, weather forecasting, petroleum, and automotive created breakthroughs impacting millions of people indirectly.

Figure 2.3 shows the evolution of Cray systems and the NEC SX-Series. The horizontal axis numbers each new system generation. The vertical axis is the year that system generation was released. One gets the feeling of an arms race between vendors with each vendor's new generation one-upping their competitors. And, in the big iron era, it was! Selling each multi-million dollar system was a big deal. Even within the Cray family engineers competed fiercely on how to one-up the previous generation. Seymour Cray himself left his current company to start a competing big iron supercomputer not just with the Cray-2 but with the Cray-3 as well. What defined a major system generation was usually a system redesign with the most recent generation of IC plus some architectural improvement. This figure shows not only pipelined vector processors, but SMM and DMM machines from these companies as well. The Cray SV-1 (1998) was the last Cray vector processor. It was a 4-processor SMM machine, which also had a feature called multi-streaming which executed vector operations by coordinating the processors to work together on vector operations.

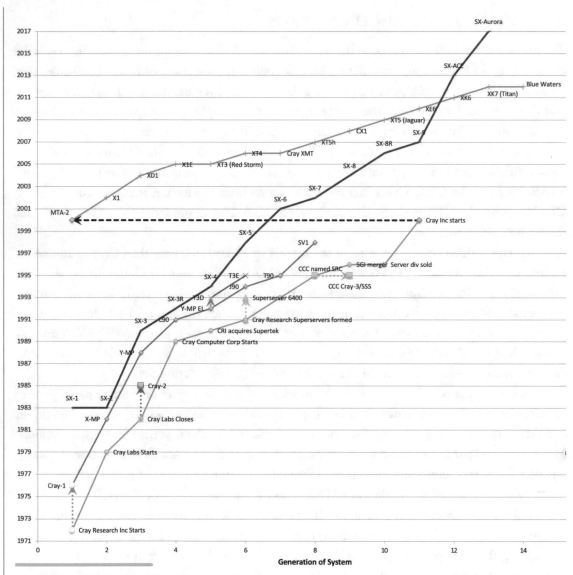

Figure 2.3: Cray and NEC SX-series machines by system generation. (The yellow Cray Research, Inc. line shows where new Cray companies were spun-off from CRI as dotted horizontal or vertical lines. Note that to reduce the space needed for the figure, the Cray system evolution restarts all the way to the left at the year 2000, the beginning of Cray, Inc.)

In Figure 2.3, we see that NEC created successful systems over several generations, as did Hitachi and Fijitsu (not shown). The evolution of all three Japanese supercomputer was strong and steady compared to the many branches of Cray's evolution. For example, the concept of multiple processors coordinating as a vector processor has been continuing for 20 years to the NEC SX-Aurora TSUBASA (2017) with 64 coordinated vector engines. An example from Hitachi is that, although cache is widely used in uniprocessors to buffer between slow memory and fast processor, during the first wave and the first part of the second wave cache was not used for vector processing leaving to the programmer and the compiler the control of locality using vector registers. As the second wave was switching to microprocessors with caches, Hitachi adopted PA-RISC microprocessors and they devised innovative architecture extensions to use register windows to preload and poststore non-contiguous array references without changing the PA-RISC's existing architecture [NNIK92].

2.4   INNOVATIVE START-UPS

The second wave was the richest period for creating parallel system companies. One can break the systems in this wave into two broad classes, mini-supercomputers and massively parallel processors (MPPs); see Sections 2.4.1 and 2.4.2, respectively. Because each of the start-ups was funded by venture capital, they had to show some differentiation from their competitors to argue that they would succeed in the marketplace. Often their differentiation was a technical innovation they brought to parallel processing, see Section 2.4.3. During the same period, it is important to note that other types of companies entered parallel processing; in particular, computer graphics companies and mainstream computer companies; see Sections 2.4.4 and 2.4.5. In the end, though, it was not too long before, "killer micros" and clusters caused a transition to the third wave of parallelism, which led to the closing of the innovative start-ups; see Section 2.4.6.

2.4.1   MINI-SUPERCOMPUTERS: CLOSE TO A CRAY AT A FRACTION OF THE PRICE

The period roughly between 1985 and 1995 was marked by widespread interest in supercomputing. The Supercomputer Conference (which has continued to this day under the simpler name of SC<number>) grew from just technical paper presentations to also include exhibits on a large show floor where each vendor, like at the Consumer Electronics Show today, tried to demonstrate some really cool applications of supercomputing, not just to the engineers of corporate customers but to the technical press as well. (For example, one year, Alliant, a company where one of the authors worked at the time, demonstrated how the Arnold Schwarzenegger skeleton running through the metal detector corridor in the movie "Total Recall" was made. You, a show attendee, could make a sequence just like Arnie had in the film. To some extent, this was a valid illustration of a new, useful technology. Alliant sold a parallel processor to a paper towel maker so that they could simulate how to make better paper towels.) Even university computing centers worked to stimulate supercomputing interest. (The University of Illinois placed a full-page ad in the *Wall Street Journal* about its expertise in parallel processing. The quite serious point was to attract corporate sponsors who would participate in joint research projects on supercomputer applications in industries that had not thought about it.) Broad interest in supercomputing also stimulated capital investment in parallel processing. Perhaps both broad interest and investment capital were a result of Cray's success. (Today's DNNP companies are in a similar position. Funded by venture capital and trying to get their differentiation into the marketplace soon enough to corner the market. At the same time they must demonstrate that deep learning can be applied in many new situations.)

The result was a period of innovative parallel computers. First there were microprocessor based shared SMM machines, including those with Uniform Memory Access (SMM/UMA). These were also known as symmetric multiprocessors (SMPs); meaning roughly the same access time was required for each processor to any memory. Second, there were Non-Uniform Memory Access systems (SMM/NUMA). They were so named because each processor had only a portion of main memory locally; and memory requests to other processor's local memory might take an order of magnitude more time than to the processor's own local memory. Third, there were cache coherent NUMA systems (ccNUMA), whose hardware guaranteed all processors would see memory changes in the same order despite concurrent accesses and updates to distributed memory. Fourth, there were Cache Only Memory Access (COMA) systems that used even more sophisticated protocols than ccNUMA to let data migrate to the best node. Finally fifth, there were Very Long Instruction Word processors (VLIW), which attempted to reorder and parallelize operations to fill memory wait times. Some of the systems in this wave incorporated no major new features, but were instruction compatible with the first wave vector supercomputers and used less expensive IC technology. As a class these systems were called mini-supercomputers. For users who did not need or could not afford a Cray, they promised computational economy, such as, half the performance of a Cray at a quarter of the cost.

Table 2.1 shows the evolution by major system generations of the parallel computer companies of this wave along with the speed of a single processor and the technology used in that generation. The architecture type of each system is also shown. If it was a hybrid parallel system, the dimension offering higher potential for speed-up is shown first. Figures 2.4(a) and 2.4(b) visualize the evolution of each company by hardware generations, horizontally, and by year, vertically. In this way one can see which companies innovated faster. The years 1982–1984 were the peak years for introduction of mini-super start-ups, seven companies opened their doors. The most successful start-ups had 4 generations of systems and the average lifetime of these companies was 12 years starting from the time they hired their first engineer.

For the most part, the innovations brought to market by these companies did not lead to the hoped-for long-term success, e.g., Alliant with their DOACROSS loop synchronization lasted only 10 years and Multiflow with their VLIW architecture, only 6. However, several of them were at the forefront of scalability at their time, e.g., Convex Exemplar, FPS T-Series, and Sequent NU-

MA-Q offered the first NUMA systems that scaled well beyond the typical 8 × 4 scaling of those days (an 8-way SMP with about 4 stage vector pipelines).

With respect to programming notations, to a large degree these companies did not attempt breakthroughs in parallel programming notations. Not being market leaders, they opted to persuade application developers to adopt a minimum of changes in their applications to get up and running quickly. And they supported as many standard programming notations as was demanded of them. With respect to parallel applications, they usually did not lead new parallel applications to market. They could not provide the highest levels of performance that application vendors could use to market their software. (By contrast, the massively parallel processor companies discussed next hoped to do just that.) However, as a whole, the mini-supercomputer companies helped parallelism penetrate a broader market. Their lower price allowed many institutions to use true parallel processors for the first time.

A promising architecture concept from the 1970s, which no company actually put into production, was dataflow [DeMi75]. There were a few dataflow processor prototypes built in labs and universities at the beginning of the second wave, such as the Electromechanical Lab Sigma-1 [YSHK85] and the Manchester Dataflow Computer [GuKW85]. These prototypes are not listed in Table 2.1 or Figure 2.2. (There are many more prototype parallel systems of all types that won't be mentioned in the book.) A key characteristic of dataflow is that operands are tagged allowing their operations to execute flexibly in parallel when the data arrives. Dataflow architectures turned out to be hard to program in some dimensions such as data locality [GPKK82]. (Some will argue that Instruction Level Parallelism (ILP) is a descendant of dataflow designs. However, tagged ILP (such as in the IBM 360/91) predated dataflow machine designs.) Only, the concept of dataflow graphs at a much coarser level has lived on as a parallel notation (Section 3.6).

Table 2.1: Mini-supercomputer start-ups. (Architecture hybrids are hyphenated with the more significant parallelism first. An asterisk in the Maximum Processors column indicates the theoretical maximum processor count. "?" indicates unknown.)

		Architecture Type	Maximum # of Processors	Technology	Processor Speed	Units of Processor Speed
(1) Alliant(Alnt) 1982–1992						
A mini-supercomputer company made multiprocessors in which each processor had vector register instructions. They migrated to the i860 RISC microprocessor and tripled the level of multiprocessor parallelism. In the end, they made a cluster DMM also.						
Alnt FX/8	1985	SMM-vector	8	Semi-custom Weitek 1064	11.8	megaflops (est)
Alnt FX/80	1988	SMM-vector	8	Semi-custom, FPU: BIT	23.5	megaflops

Alnt FX/2800	1990	SMM-vector	28	Intel i860	40	megaflops
Alnt Campus/800	1990	DMM-SMM	192	Intel i860	40	megaflops

(2) Convex(Cvex) 1982–1995

A mini-supercomputer company that also made vector multiprocessors that were generally faster than Alliant's. They used higher performance logic such as ECL and GaAs. They migrated to NUMA MPPs with RISC microprocessors.

Cvex C-1	1985	Vector	1	CMOS semi-custom, FPU: ECL	20	megaflops
Cvex C-2	1988	Vector-SMM	4	CMOS semi-custom, FPU: ECL	50	megaflops
Cvex C-3	1991	Vector-SMM	8	GaAs	50 to 240	megaflops
Cvex Exemplar SPP-1600	1994	SMM-NUMA	128	HP PA-7200	140	MHz

(3) FPS Inc 1970–1991

FPS predates Cray but later generations are second wave. (Their first system used magnetic core memory!) Mainly they made VLIW vector accelerators for minicomputers and mainframes. They migrated to MPP, first trying DMM, and then NUMA.

FPS AP-120B	1976	Vector	1	TTL Core memory	12	megaflops
FPS-264	1981	Vector	1	ECL	60	megaflops
FPS T-Series	1986	DMM	16384*	INMOS Transputer; Weitek	16	megaflops
FPS Model 500	1988	SMM	8	Sun SPARC/ BIT B5000		

(4) Encore(Encr) 1983–2002

Encore first made unsophisticated SMM multi-microprocessors systems, switching to faster microprocessors when needed. Later they migrated to a DMM systems with SMM nodes.

Encr Multimax	1985	SMM	20	National Semi. NS32032	0.7	mips
Encore-91	1991	SMM	40	Motorola 88000	25	MHz

Encr Infinity R/T300	1994	SMM-DMM	2000*	Alpha 21064	200	MHz

(5) Sequent(Seq) 1983–1999

Sequent was similar to Encore. Except that they tried migrating to NUMA.

Seq Balance 21000	1986	SMM/UMA	30	National Semi. NS32032	0.7	mips
Seq Symmetry 2000	1987	SMM/UMA	30	Intel 80386	40	MHz
Seq NU-MA-Q	1996	SMM/ NUMA	250*	Intel Pentium Pro	200	MHz

(6) Celerity(Cel) 1983–1988

Celerity was a short-lived mini-supercomputer company which like Alliant and Convex offered vector processor SMM systems. They attempted to switch to faster ECL logic and ran into financial difficulty.

Cel 6000	1985	SMM-vector	8	BIT and NCR/32	33	MHz

(7) Supertek(Stek) 198–1994

Supertek's business plan was to be a low-cost mini-supercomputer binary compatible with the Cray X-MP but slower. They succeeded at this for almost ten years.

Stek S-1	1989	Vector-SMM	Cray x-mp	CMOS	?	
Stek S-2	1992	Vector-SMM	Cray y-mp	CMOS?	?	

(8)Cydrome(Cyd) 1984–1988

Cydrome was short-lived and made a VLIW multi-functional unit processor. This allowed them to speedup some crucial applications that that did not vectorize.

Cyd Cydra-5	1987	VLIW		ECL	25	MHz

(9) Multiflow(Mflw) 1984–1990

Multiflow architecture was similar to Cydrome but more successful in business. The VLIW machines needed and had more sophisticated compilers to keep their multiple functional units busy.

Mflw Trace14/300	1987	VLIW	FPU expandable	CMOS	?	

(10) Denelcor(Dcor) 1974–1985

Denelcor preceeded the VLIW systems at attempting to speedup applications that didn't vectorize. But, they took the approach of executing many threads simultaneously (SMT). They also provided full-empty bits on memory words to ease syncrohonization.

Dcor HEP	1982	Multi-threaded	4	ECL	10	mips
(11) Tera Computer(Tara/Cray) 1987–2009						
Denelcor went out of business very early in the second wave. But, the concepts (and some of the engineers) started over again in Tera Computer with a little better success. Eventually, there were not enough applications in their domain to survive.						
Tera MTA-1	1997	Multi-threaded	4	GaAs	780	megaflops (est)
Cray MTA-2	2003	Multi-threaded	4	CMOS	~ Cray T90	
Cray MTA-3	2006	Multi-threaded		CMOS?		

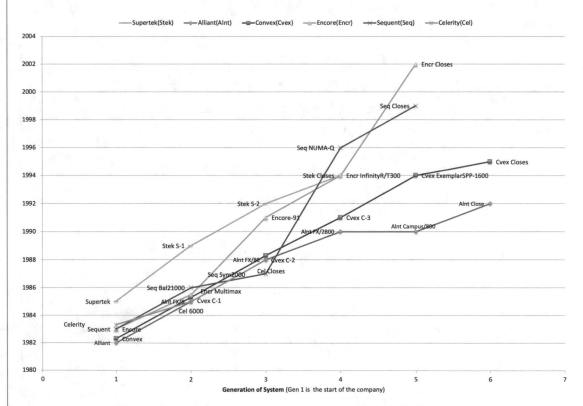

(a) Timelines for SMM, SMM-vector, and vector-SMM start-ups

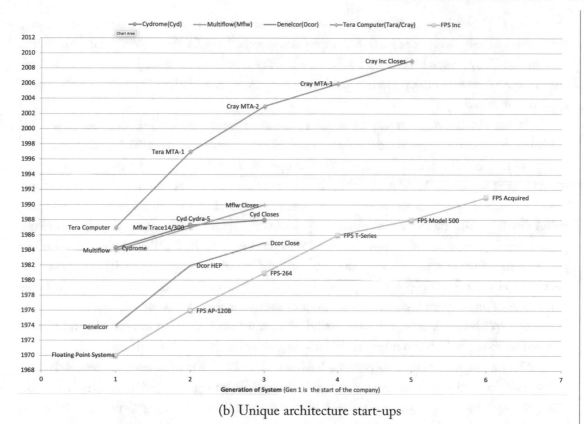

(b) Unique architecture start-ups

Figure 2.4: Mini-supercomputers and unique architecture start-ups by system generation.

2.4.2   MASSIVELY PARALLEL PROCESSORS

A second class of machines in the second wave consisted of systems with the capability of scaling to a large number of processors (typically more than 100). They were known as MPPs. These systems used either simplified custom processors or off-the-shelf microprocessors. They were SIMD, SMM/NUMA, or DMM machines. The evolution of this wave is depicted in Figure 2.5 using the same scheme as Figures 2.3 and 2.4(a) and (b). Whether mini-supers or MPPs, it was hard for custom processor systems to keep up with the pace of microprocessor evolution and the microprocessor-based companies seemed to stay in business longer. Perhaps they were more capital efficient, not having to develop their own processors or compilers.

Table 2.2 shows the major MPP based start-up companies either SIMD or MIMD. As with the mini-supercomputer companies, the years 1982–1984 were the peak years for introducing MPPs. Also included in the table for completeness are the Goodyear MPP, the ICL DAP, and the BBN Butterfly. They appeared earlier than the MPP system wave. The maximum number of processors is shown in the table together with the interconnection network characteristics. Since the number of processors configurable in a system was sometimes far larger than the largest system delivered, the largest system delivered is listed when it has been documented. There are many different ways to measure peak performance of these systems, especially SIMD systems. Thus, a metric appropriate to each system is shown.

At the same time, it would be useful to compare performance. One wonders how 1-bit SIMD array processors would perform compared to microprocessor-based MPPs. One comparison could be a SIMD array doing floating-point operations. Batcher [Batc80] gives us a few data points for the Goodyear MPP performing element-by-element array operations. From this data we estimate that a single precision floating-point add took around 380 instructions and a single precision floating-point multiply took around 750 instructions, including array manipulations. Thus, performing floating-point multiply-adds, the 16K + processors in the MPP could have roughly the performance of 16 microprocessors of the same period. On the other hand, early SIMD arrays were never intended for floating-point. They were designed for applications that favor pixel or small precision integer array operations, such as image processing. Batcher also reports 12-bit integer array operations. With this data we estimate around 37 instructions per 12-bit integer add and around 180 instructions per 12 bit multiply, again including array manipulation overhead.

Moving forward in time to 1990, each Maspar MP-1 PE was much more powerful. It contained a single precision floating-point unit as well as 48 32-bit registers [Nick90]. (There were 32 PEs per chip.) The 16K PE system's peak performance was listed as 1.2 gigaflops. The MP-2 was listed at 6.3 single precision gigaflops. (Its price was said to be comparable to a minicomputer.) [FiDo91] reports the applications ported from the Goodyear MPP to the MP-1 including: a finite element method, neural nets, and of course image processing. Performing the standard JPEG image compression algorithm the 16K PE MP-1 reached more than 25 times the speed of a 28 mips SPARC 2, a typical microprocessor at that time [CoDe93].

Moving forward in time again, recent DNNPs, listed in Table 5.1(b), frequently contain SIMD arrays with 8-bit PEs for inferencing. Most have designed-in single precision floating-point support for learning.

SIMD system start-ups struggled for several years to get broad adoption by productizing applications that had very exciting peak performance as prototypes [TuRo88]. Generally, applications had to face the inefficiency of operating part of the time with less than highly parallel kernels. SIMDs also faced the obstacle of requiring the use of their unique programming model supported by the array notations (see Section 3.1). Finally, the SIMD system companies had shut their doors by the mid 1990s. A few concepts of these designs live on in today's clusters and GPUs. Internally GPUs do not use many hop mesh, hypercube, or fat tree interconnections; they use high degrees of SMT to allow bus memory accesses to clear. Several DNNP chips on the other hand are using limited interconnection SIMD architectures because of the regular nature of their computations. See for example, Mythic's 2D array and Groq's TSP systolic array, touched on in Section 5.2. They also provide libraries of primitives that allow the developer to program using macrodataflow operations.

Table 2.2: MPP start-ups ("?" indicates unknown)							
		Primary Programming Notations	**Max Processors**	**Interconnect**	**Technology**	**Speed**	**Units (whole system)**
(12) ICL 1974–1986							
The DAP appeared before the second wave. It is included for comparison with Thinking Machines and nCube.							
ICL DAP	1979	SIMD	64x64 1-bit	4 neighbor mesh	MSI, LSI	160	microsec, 20-bit multiply
(13) BBN and ACI 1978–1991							
The Butterfly is the first MPP NUMA system. It was a linear array with a multistage crossbar switch. This was unique.							
BBN Butterfly GP-1000	1981	SMM/ NUMA	128(256)	NUMA crossbar	Motorola 68020	5	mips
ACI TC-2000	1989	SMM/ NUMA	512	NUMA crossbar	Motorola MC88100	33	MHz
(14) Goodyear (Gdyr) 1972–1991							
Both the STARAN and the MPP were SIMD arrays that appeared before wave two. The STARAN was a linear array that implement content addressable memory using the shuffle interconnection. The MPP was a 2D array similar to the DAP.							
Gdyr STARAN	1972	SIMD	4x256 1-bit	shuffle	?	40	mips, whole system
Gdyr MPP Started	1977	SIMD	128x128 1bit	4 neighbor mesh	?	910	mips, 12-bit array multiply
Gdyr MPP Delivered	1983						
(15) MasPar 1987–1996							
The MasPar systems were 2D SIMD arrays of one bit processors. However, compared to the DAP and the Goodyear MPP, each PE contained floating-point units and more memory. The mesh connections were diagonal as well as horizontal and vertical.							
MasPar MP-1	1990	SIMD	128x128 4-bit w/fltpt	8 neighbor mesh, external xbar	custom CMOS	2,600	mips, whole system
MasPar MP-2	1992	SIMD	128x128 32-bit w/fltpt	8 neighbor mesh, external xbar	custom CMOS	68,000	mips, whole system
(16) Thinking Machines (TMC) 1983–1994							
Each generation of Connection Machine (CM) was substantially different. The CM-1 was a strict 2D SIMD array. CM-2 had an attached floating-point accelerator for every 32 processors. The CM-5 switched to a microprocessor MPP with a fat-tree.							
TMC CM-1	1985	SIMD	64K 1-bit	hypercube	custom chip		?

TMC CM-2	1987	SIMD	64K 1-bit	hypercube	custom chip, plus Weitek float-ing-point unit	33	MHz (Weitek)
TMC CM-5E	1991	DMM	1024	fat-tree	Sun Super-SPARC	130	gigaflops measured, 1024 cores

(17) nCUBE 1983–2005

nCUBE made DMM clusters relatively early. They made custom microprocessors with 64bit floating-point and communication instructions to be able to put 64 nodes on one board. By the Media CUBE4 generation they switched to Intel microprocessors.

nCUBE 10	1985	DMM	1024 32-bit	hypercube	custom chip		
nCUBE-2	1989	DMM	1024 32-bit	hypercube	custom chip	1.91	gigaflops measured
nCUBE-3	1995	DMM	64K 64-bit		custom chip	6.5	teraflops max
nCUBE Media CUBE4	1999	DMM			IA32 server		?

(18) CMU Warp 1984–1994

The Warp processor was the first systolic array system. Eventually Intel won a production contract and built a custom VLIW microprocessor which allowed high-speed communication although the torus was not totally syn-cronous.

CMU/GE WARP	1986	Systolic	10 expand-able	systolic linear	?	10	megaflops
CMU/Intel iWarp	1990	Systolic	10 expand-able	systolic torus	custom chip	?	

(19) Intel iPSC 1984–2000

Like nCUBE, the iPSC was a hypercube cluster. But, it used standard Intel microprocessors. This led to the Intel Paragon which switched to a mesh connection of i860 processors that reached the top of the TOP500.

Intel iPSC/1	1985	DMM	128	hypercube	Intel 80286, 80287	12	MHz (80287)
Intel iPSC/2	1987	DMM	128	hypercube	Intel 80386, 80387	33	MHz (80387)
Intel iPSC/860	1990	DMM	128	hypercube	Intel i860	50	MHz
Intel Paragon XP/S	1993	DMM	2048	4 neighbor mesh	Intel i861	50	MHz

(20) Kendall Square Research (KSR) 1986–1994

Later in the second wave, KSR was the first production COMA processor. They built their own VLIW proces-sor to achieve high performance 64bit floating-point and the fat-tree was a hierarchy of rings.

KSR KSR1	1991	SMM/ COMA	1088	COMA fat-tree	CMOS	40	megaflops
KSR KSR2	1992	SMM/ COMA	5000	COMA fat-tree	CMOS?	80	megaflops

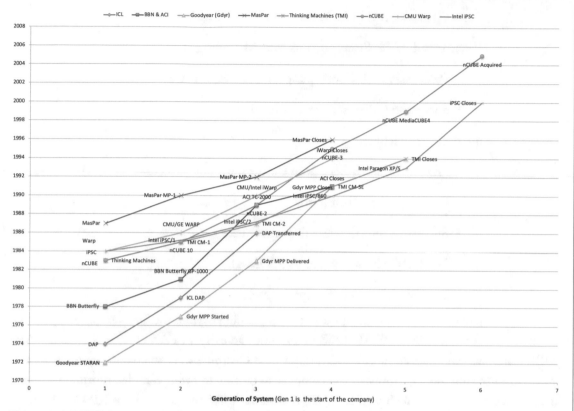

Figure 2.5: MPP start-ups by system generation.

2.4.3 THE INNOVATIONS OF THE SECOND WAVE.

The second wave is marked by the interesting parallel architecture features the start-up companies brought to the field. Here are the parallel features introduced by the second wave sorted into three bins: (Only a few predecessors and successors are shown.)

- New parallel processor features:

  ○ Multiflow and Cydrome: **Very Long Instruction Word** (VLIW)

    - Predecessor: FPS AP-120B

    - Successor: Some DNNPs, such as Groq and Habana, Section 5.2.

- New parallel threading features:

  ○ Alliant: **DOACROSS** synchronization of multiprocessors

- • Predecessor: Cray X-MP synchronization registers

  ○ Denelcor and MTA: **Simultaneous Multithreading** (SMT)

  - • Predecessor: CDC6600 peripheral processor barrel

  - • Successor: GPUs, such as Nvidia SIMT

- • New memory architectures:

  ○ Convex Exemplar, and others: **Non Uniform Memory Access** (NUMA)

  - • Predecessor: BBN Butterfly

  - • Successor: Intel QuickPath Interconnect

  ○ KSR: **Cache Only Memory Access** (COMA)

  - • Successor: Sun Wildfire (prototype).

VLIW [FERN84] was an attempt to extend pipelining beyond the limits imposed by conditionals and short loops. The FPS AP-120B of 1976 was an early VLIW machine. Unlike the AP-120B, which was an attached processor, Multiflow and Cydrome were standalone systems. VLIW emphasized scheduling algorithms in the compiler to be able to avoid complex runtime hardware in the processor such as branch prediction. (In that sense, it had the same goal as RISC.) However, VLIW systems weren't able to push the pipelining speed-up limit far enough to be called scalable in a period when there were several other promising ways to achieve scalability. Still several applications found VLIW faster than conventional vector architectures. Today, some of the deep neural net processors are re-introducing VLIW architecture for other purposes. Where Multiflow sought to speedup irregular applications, DNNPs seek to tune well-structured matrix operations.

DOACROSS [KDLS86] loop synchronization hardware was an interesting attempt to speedup loops that had a limited number of dependences. It consisted of improvements to dedicated multiprocessor synchronization registers, such as those in the Cray X-MP. Going back to machines like the Burroughs FMP prototype, it was hoped that reduction operations that are frequently only a small part of a large loop, could be executed sequentially and be hidden behind parallelism in the rest of the loop. Today two alternatives are used. If there are nested parallel loops, parallelize a loop other than the loop that contains dependences. Or, isolate dependences into collective or reduction operations where the strategy is to provide a kernel that is tuned for each system configuration.

Since the latency to main memory on HPC systems is high, memory latency hiding mechanisms were sought to compensate for the growing disparity between fast processors and slower memories. The most influential such mechanisms was Simultaneous Multithreading (SMT). SMT appeared in second wave systems, such as the Denelcor HEP [Smit78] and the Tera Computer

MTA [SCBM98], which contained many register sets to quickly swap the register set of a thread that stalls to the register set of possibly many that are ready to execute. For example, a four processor MTA-1 supported 512 threads. (Denelcor and MTA also introduced full and empty bits on memory words for synchronization.) Multiple register sets for multithreading is a concept that lives beyond the HEP and the MTA. In many GPUs, it is part of Single Instruction Multiple Thread (SIMT). And with fewer register sets in microprocessors, hyperthreading is used, which may only support two threads. (Dating back to the CDC6600 [Thor70], the world's fastest computer between 1964 and 1969, the peripheral processing unit could compensate for the latency of IO devices compared to the CPU. The "register barrel" was 10 registers sets that swap on every clock cycle whether stalled or ready.)

Multithreaded processor techniques do hide the latency to main memory as seen by the thread that is blocked but they do not make the stalled thread run any faster. Thus, other mechanisms to deal with latency were sought. Cache Only Memory Architecture (COMA) [DaTo99] was built by KSR Corp. They developed an MPP in which data from other nodes migrated to the local memory of the node using the data; it was called "Allcache". (NUMA systems have a portion of memory allocated statically to each node.) In COMA, if a memory reference was already in the local node, performance was good. If not, the whole memory was searched by the hardware memory protocol to bring the access to the node accessing it. Given that the cached line was now node local, no more network traversals were necessary. Thus, COMA could offer improved latency in MPPs on average. However, at this point in time commodity microprocessors, which could be integrated into a NUMA architecture, were faster than KSR's custom multi-chip processor, KSR went out of business. After that a few other prototypes of COMA systems appeared, e.g., the Sun Wildfire. But as far as we can tell, none were put into production.

Many of the second wave companies saw they would have to transition to MPP microprocessors architectures. How could they scale-up and at the same time have an integrated easily programmed system? NUMA appeared to be the answer. NUMA systems are broadly defined as microprocessor-based systems with limited interconnection topology in which the virtual memory architecture was integrated with the interconnection networks. Thus, MPPs could appear to share main memory across all nodes and message passing to get the data was hidden from the programmer by the system. SMM compatibility and notations could be kept. This was true except that frequently applications needed to be tuned to utilize the NUMA protocol more efficiently and each NUMA system had differences in protocol, interconnection, and memory. NUMA microprocessor systems, although not a perfect solution, generally won over KSR's COMA [SJGH93]. But given one had to do the tuning, NUMA eventually also lost to simpler, less costly DMM machines as DMM notations, e.g., message passing, diffused into the skills of many developers.

Although the SMM was not new by the second wave, this section would not be complete if we did not mention its importance to parallel programming at that time. An SMM architecture in which each thread can execute a different control path makes it inherently easier to parallelize new applications and developing more parallel applications was important then. In contrast, certainly on SIMD machines, and sometimes on vector machines, one should conceive and code the application anew for efficient operation. In addition, comparing an SMM to a DMM system, on an SMM system it is not necessary to explicitly program communication. Hence, the second wave mini-super-computers almost all opted for a strong multiprocessor dimension. Gradually, over 20 years or more of diffusion of more sophisticated parallel programming techniques, many applications can run effectively on architectures that are harder to program than an SMM machine and, in the today's third-wave high-end systems, SMM multiprocessing, albeit ubiquitous, is not the primary dimension. Today, adapting algorithms to restricted, simpler, but faster, hardware can avoid the expense of a globally shared memory machine. For example, BLAST, a gene matching algorithm discussed in Section 4.4, was first parallelized on an SMM machine during the second wave because the loop selecting each gene could be converted quite simply from sequential to parallel loop form with the addition of a few reduction operations. Later, the BLAST was adapted to run on a DMM machine using MPI. It has even been adapted to run on GPUs. In the final analysis, one tends to trade-off the price of added hardware for greater programming complexity.

In summary, the second wave showed remarkable experimentation by start-up companies and, more broad than the particular examples above, they put into use the general parallelism concepts found today: multiprocessors with several varieties of shared memory architecture with SMM notations, and limited topology interconnection network-based DMMs using message passing notations. However, don't forget the examples above. We will see that several of them will come back.

2.4.4   GRAPHICS IMPACT ON PARALLEL PROCESSING

Before moving to the third wave, this and the next two subsections note a few topics that supported the rise of parallel processing. Particularly noteworthy is that the first large end-user market to utilize parallel processing was computer graphics. Computer games and graphic workstations were key proof points for parallelism. There was a brief period when venture capital funded multiprocessor visualization workstation companies such as Stellar, Ardent, and then their merged company Stardent. Multiprocessor workstations from pioneering companies such as SGI, Silicon Graphics Inc., and Stardent led to commodity SMM workstations from companies like Digital, HP, and IBM, which dropped the visualization features. Many interesting applications were parallelized for scientific and engineering use on these workstations to the benefit of the multicore PCs that came later. With respect to machine design principles though, there was not much new in these workstations. Finally, single-chip GPUs could be added into PCs and several years later integrated into the processors on laptops with very little extra cost.

A curious example of graphics impact on parallel processing is the ES-1 from Evans and Sutherland. They pioneered graphic systems in the 1970s and were prosperous in the 1980s and 1990s. They produced very high-performance graphic systems for special applications such as flight simulators and digital theaters. These systems turned out to be so close to general-purpose computers that Evans and Sutherland attempted their own supercomputer called the ES-1 in the late 1980s. It was a multiprocessor with impressive benchmark performance. Its downfall was that applications other than graphics could create high memory contention.

The biggest impact on parallelism in engineering and science among graphics companies was SGI. Founded in 1981, they originally produced just graphics workstations. Then, as well as SMM workstations, they successfully produced many different parallel systems from super-mini-computers to NUMA systems. In fact, SGI's critical mass in the market probably hastened the demise of some of the mini-supercomputer companies. Finally, in 2009 SGI disappeared for multiple reasons: as graphics expertise diffused and GPUs reached commodity markets, graphic performances was not the differentiator it had been when they were founded; supporting their proprietary RISC architecture, called MIPS, was costly in the face of commodity microprocessors; consequently, they moved to Intel Itanium, which itself ran into trouble; finally, their strategy of fully integrated systems when parallelism was moving into COTS clusters had the same problems for them as other second wave companies.

As noted throughout this book, GPUs have had a major impact on parallel processing. The first GPU with enough complexity to be a capable parallel processor, i.e., capable of 3D polygon rendering, appears to be the Namco System 21 created in 1988 for arcade games. It was a system with 2 Motorola 68000s and 4 Texas Instruments DSPs. Since then GPU parallelism has grown to thousands of small cores on an IC.

Applying GPUs for GPGPU faced similar challenges to those faced by Stardent, Evans and Sutherland, and SGI. But, this direction has been more successful. First, they are inexpensive, which the others weren't. Also, parallel notations for GPUs have been more up to the challenge for three reasons: high-level languages like CUDA, carefully tuned kernels provided to and by the community, and an investment to accelerate diffusion of GPGPU by funding university research programs and posting GPU techniques and algorithms on the web.

2.4.5   PARALLEL PROCESSING FROM MAINSTREAM COMPANIES

A trend toward general parallelism technologies was started later in the second wave by many of the mainstream computer companies (OEMs) who advanced parallel processing as a portion of their broad product lines in two ways. First, they successfully commoditized less aggressive parallel architectures. And second, some of them directly supported start-ups for a time. Why did they do this? Later in the second wave, reciprocal arrangements between the mainstream system providers and the parallel processing companies could be a win-win. The parallel processing companies saw they were losing the performance battle with microprocessors and that they could no longer afford the time or cost to design their own processors, which would be slower than the microprocessors by the time they were released. At the same time, the mainstream companies were looking at the threatening progress in x86 architecture and they were persuaded they could make their own faster microprocessors based on RISC architecture. Realizing they needed to get design wins outside their own systems to compete with Intel, they saw the parallel processing companies as potential adopters of their RISC architecture. Figure 2.6 shows OEMs that produced RISC systems, what parallel system companies adopted which RISC architecture, and some of the parallel systems they went into. Into the third wave, OEMs started merging and some of these acquisitions are also shown. (There are more RISC architectures, adoptions, and acquisitions not shown.)

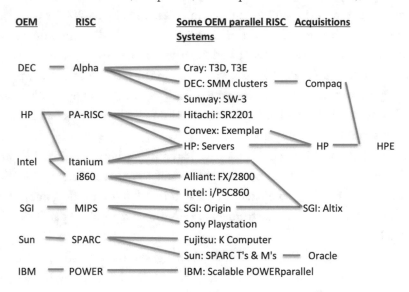

Figure 2.6: RISC architectures in parallel systems. (Itanium is included because it evolved from RISC.)

For some examples of relationships between RISC systems and parallelism companies:

- Digital Equipment Corporation (DEC): Cray adopted DEC's Alpha architecture for their T3D, T3E series. DEC, by working with Cray on the T3D and T3E, is said have been able to double Alpha RISC performance.

- HP: Hitachi, leaving their vector architecture to develop an MPP system, supported HP's PA-RISC for their supercomputers and then extended PA-RISC by developing a system of register windows for vector processing.

- SGI: First made their own RISC architecture, MIPS. Later they partnered with Intel on Itanium NUMA systems at one point own the second spot on the TOP500, Columbia.

- Sun: Fujitsu adopted the SPARC architecture and from 1995 manufactured SPARC multiprocessor workstations and servers, e.g., PRIMEHPC series and the K Computer that was number one on the TOP500 in 2010.

2.4.6   THE END OF THE SECOND WAVE: SUPERCOMPUTERS MOVE TO MICROPROCESSORS, DMM, AND OPEN SOURCE

The second wave of parallelism start-ups were overtaken by a new wave in less than a decade in a sequence of steps that is as amazing as many TV series.

- First, around 1980, the BBN Butterfly was an idea ahead of its time, a parallel processor made from over 100 microprocessors, i.e., an MPP, connected by a high-speed interconnection fabric [LeSB88]. Unfortunately, its interconnection network turned out to be slower than the microprocessors that they used, MC68000s, which were themselves slower than the custom processors of early second wave companies. It was successfully employed in only a few production applications.

- Second, early second wave companies, such as Convex and Alliant, appeared to be doing well with custom processor mini-supercomputers. However, all was not well as a business. In those days, a system buyer wanting more application performance had a choice of going with a new moderately parallel Vector-SMM architecture where it took at least a year to learn and apply the parallelism to their applications or they could wait the mythical 18 months for Moore's Law to produce microprocessors that were nearly twice as dense and fast with no program changes. Many pragmatic buyers chose to wait. This was a problem, since, in the start-up business model, venture capitalists had strict financial requirements that refused granting additional rounds of funding if growth slowed.

- Third, the packaging of ICs as density increased meant that any processor implemented in more than one IC was almost certainly slower than a microprocessor. Thus, this led to the "killer micro," which appeared by 1990 [Broo90]. Companies like Alliant and Convex tried get back on track by converting to RISC microprocessors, which were faster and, by going to a single-chip processor, the company could double, maybe even quadruple, the number of processors on a board. In addition, new interconnection network hardware was being used in datacenters bringing costs down. At this juncture it looked like BBN was right, an MPP system using microprocessors, assisted with proprietary inconnection protocols, would allow for shared memory and easy programming … NUMA was the way to go.

- Fourth, additional systems, like nCube and i/PSC, based on the Caltech Cosmic Cube experience proposed that fast microprocessors and inexpensive networks could be nearly as good as other parallel systems of the time without NUMA hardware. Thus, DMM message-passing systems got a foothold. (Not knowing this would bring about the demise of these systems.)

- Fifth, a few years later, everyone started seeing that RISC was not much faster than CISC x86 at any point. And that x86 generations were coming faster than any one RISC provider could go through the design-fab cycle. So ended the second wave and so entered the third wave, with the Beowulf cluster. (Only a few RISC architectures would survive, IBM continues producing their POWER, Performance Optimization With Enhanced RISC, and ARM, although under the radar for several years, it is having a resurgence.)

One should note that the second wave was crippled for software reasons as well. Although all these start-ups ran some variation of UNIX, the "Linux Revolution" meant that system managers could reject the economics of buying a complete system solution with continuing hardware and software support. This upset the second wave business plans, which were formed before the Linux revolution.

With respect to programming notations, with exception of message passing, there were as least as many parallel notations at the beginning of the second wave as today so that becoming a parallelism expert is not much easier today. However, today the expertise has diffused into many applications, application specific libraries, and cluster management tools so that fewer users need black belt skills. By diffusion we mean that a major asset of the start-up companies of the second wave was their engineers, both hardware and software. Skilled in practical parallelism, they went on to other companies to develop the parallelism we see today.

2.5   CLUSTERS OF MICROPROCESSORS AND COMMODITY PARALLELISM

The third wave, DMM clusters and commodity parallelism, started before the end of the second wave. The third wave is not as active as the second wave was with parallel system companies coming and going, but the flexibility and choices of commodity parallel processing hardware is richer than ever and DMM systems have grown into true supercomputers. There are several choices of parallel processing ICs that can be used in supercomputer nodes and now in ever-smaller parallel systems for personal systems and Systems-On-a-Chip (SoCs) in embedded devices. Today, parallel processing has found its way into nearly every computer system from the largest to the smallest for applications to adopt to varying degrees. As such, despite all the comings and goings in the first and second waves, parallelism has succeeded.

This section contains two groups of three subsections each. Supercomputer clusters will be discussed in the first three subsections: 2.5.1 discusses the DMM clusters from Beowulf systems to the TOP500 clusters of today; 2.5.2 looks forward to the challenges ahead as clusters continue to grow; and 2.5.3 shows that many systems today are hybrid parallel systems combining some form of vector processor, as well as SMM, and DMM parallelism.

The second group is an ingredients view of the third wave: 2.5.4 covers how interconnection networks that started in the second wave evolved into the network fabrics of today's supercomputer clusters; 2.5.5 covers IC architectures used in parallel processing: multicores, GPUs, ASICs, FPGAs, and SoCs; and 2.5.5 touches on how networks of workstations and cloud computing can "virtualize" parallel processing resources.

2.5.1   EMERGENCE AND TRIUMPH OF DMM CLUSTERS

There are two types of DMM machines. *Ordinary clusters*, or simply *clusters*, are DMM machines made with Commercial Off The Shelf (COTS) microprocessor boards and switches. Many of the systems in the TOP500 today are ordinary clusters. There are also *integrated DMM* machines, which are in the minority of the TOP500 (see Figure 2.7) and use customized components, typically the interconnection network, and perhaps for I/O. They may also have optimized development and operating software. All intended to achieve petascale or exascale computing. Examples of integrated DMM machines include the Sunway TaihuLight System consisting of a Sunway custom network connecting manycore Sunway SW26010 processors and the Cray XC50 consisting of Cray's proprietary "Aries" network connecting Intel Xeon processors, Cavium ThunderX2 processors, and/or NVIDIA Tesla P100 GPUs. An example of an ordinary cluster is IBM's Summit consisting of nodes containing two IBM POWER9 CPUs and six Nvidia Tesla GPUs connected via a dual-rail Mellanox EDR InfiniBand. As illustrated by these examples, both types of DMM machines today have nodes that are multicore microprocessors and could include accelerators, usually GPUs as a result of GPGPU research [PhFe05, OLGH07].

Clusters arrived quickly; Gordon Bell described Beowulf's [SSBD95] growth in one paragraph in 2002:

> *"The Beowulf Project was based on commodity technology. The project was started in 1993 based on NASA's requirement for a one gigaflop workstation costing under $50,000. The recipe for building your own Beowulf is contained in 'How to Build a Beowulf'* [Ster99]. *A 16 node, $40,000 cluster built from Intel 486 computers ran in 1994. In 1997, a Beowulf*

cluster won the Gordon Bell Prize for performance/price. By 2000, several thousand node computers were operating. In December 2000, 28 Beowulfs were in the TOP500 supercomputer centers and the Beowulf population is estimated to be several thousand since any technical high school can buy and assemble its own from standard parts." [Bell02]

Figure 2.7 shows how, over the course of 25 years, DMM machines, encompassing both integrated DMM machines and ordinary clusters, grew to constitute all of the registered TOP500 systems. The clusters targeting exaflop performance will follow this trend, as discussed in Chapter 5.

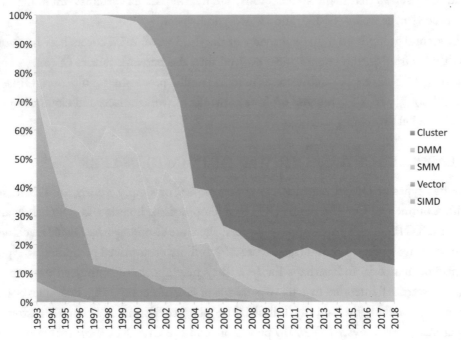

Figure 2.7: Distribution of the TOP500 systems by architecture type over time. (The TOP500 organization database has a somewhat different set of labels than those used in this figure. Here: Cluster = Ordinary clusters with: single processor, SMM, or accelerated nodes; DMM = Integrated DMM machines with customized nodes or interconnects; SMM = Shared Memory Model machines: simple SMMs or NUMA-SMMs; Vector = Vector pipelined machines: single processor, SMM-vector machines, or NUMA-vector machines; and SIMD = SIMD machines such as the Goodyear MPP.)

Note that the emergence of cluster technology has caused a globalization of parallel processing. Where the first and second waves were mostly led by the U.S., cluster technology has enabled centers of parallelism excellence around the world.

- The Japanese Fugaku cluster took over the top position from the Summit cluster in June 2020. Before that, Japan held the top position several times: the K Computer, the Earth Simulator, the Hitachi SR2201, the Numerical Wind Tunnel, the NEC SX-3/44, and the Fijitsu VP2600/10.

- China has also held the top position several times with TaihuLight and Tianhe-2A. China now, in 2020, has 226 systems in the TOP500 compared to only 114 in the U.S.

- Europe is well represented in 2020 with three systems in the top 10 located in Italy (2) and Switzerland (1). In terms of the number of systems per country, France has 18 systems in the TOP500 (4th most supercomputers) and Germany 16 (5th most). Followed in rank order by Netherlands (15), Ireland (14), and the United Kingdom (10).

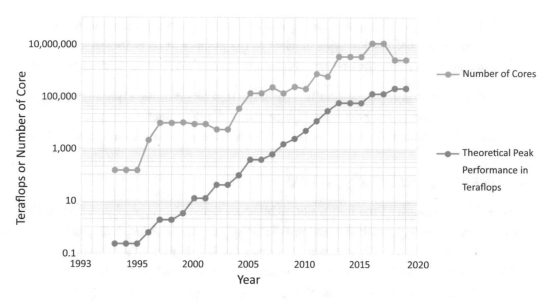

Figure 2.8: The peak teraflops for the number-one system in the TOP500 ranking each year.

2.5.2 INCREASING THE NUMBER OF PROCESSORS IN A CLUSTER RAISES CHALLENGES

As Figure 2.8 shows, at the high end, as performance has increased, so has the number of cores. The challenges that arise as the scale of parallelism continues to increase have been discussed in the literature (see for example [Snir13]). Challenges of delivering very large-scale parallelism in applications include [SaHS09] the following.

- Scheduling dynamic parallelism with fine-grained tasks: Scheduling in a coarse-grained parallelism environment is relatively easy. Here we refer to applications that have phases requiring fine-grained parallelism that can't be prescheduled. If the fine-grained tasks can be prescheduled, even to a limited degree, it may be possible to combine them into coarse-grained tasks, doing reductions locally before globally is a classic example. If prescheduling is not an option, fine-grained synchronization can take thousands of processor cycles, which can easily take longer than a fine-grained task.

- Distribution and co-location of tasks and data: Data placement on large numbers of cores must be a dominant consideration. As a significant aside, data placement is most often implicit in most of today's programming notations and to be reasoned about it must to be documented in the code.

- Collective and point-to-point synchronization with dynamic parallelism: Managing to keep all synchronization types efficient as the number of cores increases is a large challenge. In addition, the best synchronization strategy may change with each dataset.

- Producer-Consumer parallelism: This challenge is to program the application to move from one phase to the next without incurring a drastic reduction in speedup as the summarized data produced by one phase is broadcast and consumed in the next phase.

2.5.3   HYBRID PARALLEL SYSTEMS

Looking separately at each of the four major architecture types: Vector pipelined, SMM, DMM, and SIMD may let an important observation fall through the cracks. Since 1980, several systems have successfully combined multiple architecture types, i.e., hybrid parallel architectures.

The first hybrid parallel systems to appear were Vector pipelined-SMM. In the first wave, in the early 1980s, starting with the Cray X-MP, groups of vector processors were organized in SMM configurations. The first programming notations that allowed combining multithreading and vector processing in one application appeared at that time, for example Cray microtasking worked with vectorization.

Late in the second wave, in the 1990s, and even today in the third wave, SMM technology became prevalent enough that DMM machines could be built with multiprocessor nodes. The ASCI White system appears to be the first large cluster of this type. It enabled for the first time a hybrid parallel programming notation with MPI for pleasingly parallel loops and OpenMP for inner loops. This combination worked well for parallelism rich applications. There is nothing specific to MPI or OpenMP here; other well-crafted programming notations allow hybrid programming. A nuance arises here that the hardware packaging can be hybrid while the programmer's view can be simpler. One example is using message passing for all parallelism, including within an SMM node. This is possible and even preferable in some applications. Also, MPPs that adopted NUMA could relatively easily use symmetric multiprocessing within the nodes.

Three-way hybrids also appeared, primarily Vector-SMM-DMM and SIMD-SMM-DMM (considering GPUs as a SIMD type). In the third wave, in the middle of the 2000–2010 decade, DMM clusters with SMM nodes could start using vector extension such as MMX or SSE or GPUs could be added to the nodes [FQKY04]. Here a programming notation such as: BLAS library for vector extensions or CUDA kernels for a GPU, could be combined with a multithreading notation, combined with MPI message passing. The latest high-end clusters continue this trend of 3-way hybrids. For example, Summit nodes have two 22-core Power 9 CPUs with 6 NVIDIA Tesla V100 GPUs and uses MPI, multithreading, and CUDA. Thus, at least at the high end, exploiting highly parallel systems fully can lead to highly sophisticated applications.

Figure 2.9 illustrates the evolution of hybrid parallel architectures. Arrows indicate when a new hybrid appeared. The names of the architectures and their hybrids are in red; the text in black shows interesting first examples of hybrids. (The colors of the bars simply distinguish one bar from another.) Among other combinations, the figure shows the following.

- Vector-SMM hybrids arrived on the scene in the period 1983–1985, e.g., the Cray-XMP, and continues in today's multicore microprocessors that all have vector instruction extensions.

- In the period 1998–2000, the Cray SV-1 created another hybrid of an SMM system containing vector processors. Recall, this type is one in which multiple SV-1 processors can work together to execute vector SIMD instructions.

- In 2004–2006, clusters of SMM nodes, i.e., DMM-SMM machines, such as the Alliant Campus and ASCI White appeared. Around 2010 DMM-SMM hybrids evolved into the 3-way hybrids described in the previous paragraph [MiVe15].

Meanwhile, interesting special architectures such as: dataflow, many-threaded, latency hiding, and systolic processors were not prevalent enough to form hybrids.

At the bottom of the figure, one can see the concurrent advances in programming notations and the major breakthroughs in the seven applications reviewed in the book.

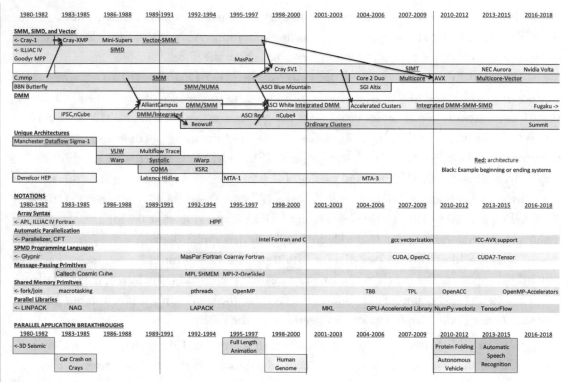

Figure 2.9: Over time, parallel architectures become more sophisticated as hybrids.

2.5.4 INTERCONNECTION NETWORK EVOLUTION

Interconnection networks have improved with each of the three waves of parallel hardware. Ideally, one would like an interconnection network to be an all-to-all crossbar switch connecting thousands of processors. But, starting with SMM vector supercomputers of the 1980s, 16 × 16 switches were a rule-of-thumb limit of scalability for high-speed switches, and 8 × 8 was most common. (In the book, the term "switch" always means crossbar switch.) (Scalability is limited by the asynchronous arbitration logic between ports, which is increasingly less deterministic as the switch grows. To avoid this, more logic may be added, which unfortunately slows the switch down.)

In the 1980s there was considerable research into scalable networks [Park80]. Two broad topological classes were used. The first is a limited topology network, such as the 2D mesh in the Distributed Array Processor (DAP) from International Computers Limited (ICL), whose topology meant that to move data from any processor to any other takes several (or many) hops through intermediate processor nodes. Second, a multi-stage network, such as the STARAN or the BBN Butterfly in which intermediate hops did not pass through processors but through internal switch nodes. So that if the switch nodes are $m \times m$ and the number of processors is n, the minimum number of intermediate nodes is log$m(n)$. The BBN Butterfly interconnect was composed of 4×4 switch elements, each with a bandwidth of 32 MB/sec. The interconnection network could accommodate 512 processors although the largest documented Butterfly system had only 128 processors.

In the second wave, SIMD and MPP systems of greater scalability used networks of several sorts, frequently: a mesh (e.g., MasPar), hypercube (e.g., iPSC), or fat-tree (e.g., SGI Altix). By the third wave, interconnection networks for cluster systems, sometimes called message passing fabrics, started to appear. It was found that message passing could be accelerated with generic switch nodes that were microprogrammed to support standards such as: in the late 1980s, HIPPI; around 2002, 10 Gigabit Ethernet, called "10GigE"; or in the early 2000s, InfiniBand. In addition, these switches could be tuned to optimize MPI directly. These could also be wired and tuned to the needs of the cluster being built and the dominant applications to be run on the cluster. Tuning for reductions operations for example.

The switch nodes for these fabrics fit into the same standard chassis as processor nodes. And, breaking the old 16×16 rule-of-thumb, 32×32, and even 64×64 switch nodes are available especially when optical links are used [KiTG05]. In addition, in each generation of the interconnection standards, the links between nodes could be increasingly wider, i.e., more parallel. For example, InfiniBand links can be "aggregated" to be 4×, 8×, or 12× wide and can be bidirectional. Thus, an interconnection network design can trade-off between ports and link aggregation.

To bring the evolution of network topology up to date, the fat-tree [Leis85] still appears to be a popular topology. And, although numerous ordinary clusters have adopted the InfiniBand standard, networks for large integrated DMM machines such as those produced by Cray increasingly use the dragonfly topology [KDSA08]. This is to the extent that dragonfly networks are no longer considered custom but an ordinary product. Dragonfly topologies are more costly because they use longer optical fiber cables and high-radix switching nodes. Driving adoption, dragonfly is especially beneficial for applications that can take advantage of minimum hop interconnects, for example seismic processing and molecular dynamics applications sometimes uses 3D FFT methods. 3D FFTs force many small messages in either the X, Y, or Z FFT transform. To avoid a bottleneck in that dimension, a low number of hops is ideal.

Here are a few examples of interconnections networks today.

- The Cray Shasta system which will be delivered to the U.S. Dept. of Energy in 2020 will contain an Aries interconnect with the dragonfly topology. It has at most three hops between any two nodes and is built with Gemini Router ASICs.

- The Aurora system anticipated in 2021 or 2022 will also have a Cray interconnect with dragonfly topology with 64 port switches running at 12.8 Tb/s.

- The Sunway TaihuLight system has a custom interconnect called the Sunway Network which connects nodes with PCI-E 3.0 which operate at 12 GB/sec and has a network diameter of 7 [Dong16].

- The Summit system is an ordinary cluster using the standard commodity interconnect in a fat tree with Mellanox InfiniBand EDR nodes.

In addition to the primary interconnection network across the cluster there are often secondary interconnection networks. The common secondary interconnects are: a system management interconnect (necessary on even the earliest clusters), a high-speed small-message interconnect for fine-grain tasks such as reductions, and an interconnection network to the IO subsystem.

It should be remarked that during this evolution, the packet routing software inside both custom and standard switch nodes has become much more sophisticated at minimizing congestion and can be tuned to various protocols, such as MPI to maximize throughput in HPC applications. For example, adaptive routing software that avoids tail latency with backpressure congestion control is a feature of the Cray interconnects just mentioned for Shasta and Aurora.

2.5.5   MULTICORES, GPUS, SOCS, FPGAS, AND ASICS

The need for parallelism is not confined to high-end systems. It is appearing increasingly in personal and embedded devices. As the scale of integration increased and clocks stopped getting faster, it was inevitable that parallel architecture would appear in most types of integrated circuits used in all types of systems. In fact, the parallel processing community noticed that types of ICs used in personal and embedded devices could sometimes provide higher density computing and consume less power than conventional parallel systems. This section looks at five types of ICs that feature parallel processing: Multicores, GPUs, SoCs, FPGAs, and ASICs.

Multicores

The number of cores frequently gets the attention when one speaks of parallelism in microprocessors. But here, the evolution of two other aspects of parallelism in microprocessors is noted.

First, SIMD processing inside each core has broadened steadily over the last 20 years. Table 2.3 shows the history of Intel and x86 SIMD instruction extensions starting from 1997 as presented in [HuHL17].

Table 2.3: Evolution of SIMD instructions in x86 [HuHL17]			
Acronym	**Name**	**Description**	**Year**
MMX	Multimedia Extensions	64-bit MMX for packed integers	1997
SSE	Streaming SIMD Extensions	128-bit XMM for floating point	1999
SSE2	Steaming SIMD Extensions 2	XMM supports doubles and integers	2001
SSE3	Streaming SIMD Extensions 3	Horizontal operations added	2004
SSSE3	Supplemental SSE3	Horizontal and data movement	2006
SSE4.1	Streaming SIMD Extensions 4.1	Extra functionality	2007
SSE4.2	Streaming SIMD Extensions 4.2	Vector string instructions	2008
AVX	Advanced Vector Extensions	256-bit registers that contain packed floating point numbers	2011
FMA	Fused Multiply Add	Fused multiply add instructions	2011
AVX2	Advanced Vector Extensions 2	Supports for packed integers	2013
AVX512	Advanced Vector Extensions 512	512-bit packed registers	2016
AVX512 VNNI	AVX512 plus Vector Neural Net Instructions	Fused instructions for packed 8- and 16-bit	2018

Second, power consumption and density is a major design factor that architects traditionally struggled to manage. Over the last 15 years, with multicore growing to 20, 30, 40, ... cores, more often than not running all the cores in parallel at their peak clock rate would exceed the power limit of the package. Therefore, architects began allowing end users to choose their operating point on the power-performance spectrum by altering the clock frequency of some cores. (Recall that in CMOS the processor speed has a major impact on power consumed: Power $\sim C * Voltage^2 * Frequency$.) Here are a few of the dynamic frequency scaling modes developed over the last 20 years (Intel names are used, but other microprocessors have similar power controls).

- Enhanced Speedstep (2005): Power state (P-State) was introduced to map requested power-level to on-chip voltage and hence frequency.

- Turbo Boost (2008–2010): With a single core running, how fast can it go. If the core is using vector mode, e.g., AVX-512, a different turbo frequency may be used.

- Speed Select (2019): Speed Select allows the operator to define several profiles of core frequencies so that the best profile can to matched with each workload.

Power management can be important for datacenter operators interested in minimizing power consumption and for end users interested in maximizing performance of their personal system. It raises several interesting questions for the user such as: How much parallelism do I really benefit from? What's the best clock rate for as many cores as are used? How long does the workload run at peak power? And, how aggressively can the trade-off be managed? Of course, if one does not manage the power and speed parameters, the microprocessor will manage them automatically with built-in thermal sensors. (Cores that are turned off are called "dark silicon" (see Section 5.7)).

Graphic Processing Units (GPUs)

As noted in Section 2.4.4, computer graphics has played an important part in parallel computation. GPUs followed Moore's Law evolving independently of microprocessors. Traditional GPUs have integrated raster processing units, texture processors with on-chip texture memory, and shader cores that contain 32-bit integer and floating-point ALUs to perform the rendering pipeline (see also Section 4.3). In 2001, it was realized that the GPU compute density was exceeding that of microprocessors. Also, GPUs' streaming architectures: loading from main memory to GPU memory, executing vector kernels, and returning the computed data to memory, made them well suited for accelerating long vector kernels in applications where data size was growing rapidly. At that time with the GeForce 3, Nvidia introduced programmable shaders, and the GPGPU programming research community researched applying GPUs to diverse applications, e.g., [WHwu11]. Out of this architectural and application change arose several programming notations for developing and interfacing GPU kernels: OpenCL, CUDA, OpenACC, OpenMP accelerator support, and others.

Initially, the cores in a GPU were controlled essentially in SIMD mode. As the number of cores increased, they were laid out into groups to place several groups on one die, each with a separate control unit so that the scheduling logic for one or multiple clusters has improved to allow increased efficiency and overlap.

GPU design is still driven by graphics performance where a major application goal is real-time ray tracing (see Section 4.3). For example, the Nvidia Turing architecture has added special-purpose cores called RT cores replacing some of the CUDA cores. Over the last few years the emergence of deep learning in applications, including computer games, has given birth to tensor product cores on Nvidia GPUs. They accelerate 16-bit floating point or 8-bit integer math to increase the effective compute density further. Hardware supported L2 cache on newer GPUs allows even more programming flexibility. Table 2.4 shows the recent evolution of Nvidia GPUs. (Notes: see also Section 5.5. Volta is comparable to Turing but designed for datacenters. The next Nvidia architecture, called Ampere, is expected in 2020 and will also have personal device, GA100, and datacenter, A100, variations.)

Table 2.4: Recent evolution of NVidia GPUs					
	Kepler (2012)	Maxwell (2014)	Pascal (2016)	Turing (2018)	Ampere (GA100) (2020)
FP/INT Cores	1536	2048	3584	4352	8192
Tensor Cores	—	—	—	8	512
FP teraflops	~3	~4.5	~12	~15	~23
Register file (KB)	48	96	96	192 partitionable	192 KB
Per SM					
L1 cache (KB)	—	—	48		
L2 cache (MB)	.5	2	~3	~6	40
Transistors (billion)	3.5	5.2	12	18	54

SoCs and MCMs

Regardless of whether Moore's Law continues, functionality and performance may continue to increase for System on a Chip (SoC) and Multi-Chip Module (MCM) devices because more sophisticated power management methods can be used and the limits are based mostly on package power. Also, parallelism inside commodity devices, personal, embedded, and Internet of Things devices, is likely to continue because they currently have less functionality and hence use less power than high-end multicore and manycore devices. An SoC may sound like a better choice than an MCM because a single-chip SoC would have faster connections than an MCM of multiple chips. However, today's MCM technology facilitates choosing the optimum die fabrication processes for distinct functions and the MCM's substrate can provide comparable lithography widths. The discussion of SoCs and MCMs continues in Chapter 5 where options for extending Moore's Law are taken up. As an example of these technology capabilities, Table 2.5 shows the growth in on-chip parallelism over three generations of the Apple iPhone. The term "performance core" refers to a core constructed with high-performance building blocks, in contrast to an "efficiency core" that implements the same functions but is constructed with smaller, low-performance, building blocks. (These are sometimes referred to as "big cores" and "little cores" respectively.) Considerable energy is saved by executing on an efficiency core normally and switching to a performance core only when there is benefit for high-performance execution. This extends battery life in a smartphone, but it is also now used in PCs and even multicore servers.

Table 2.5: SoC components in several generations of Apple iPhone			
	iPhone 1 (2007)	iPhone 4 (2011)	iPhone Xs with Apple A12 Bionic (2018)
CPU	Single core (620 MHz)	Dual core (1 GHz)	2 Performance + 4 Efficiency cores (2.5 GHz)
GPU	103 MHz	200 MHz	4 GPU cores + 8 cores Neural Net
DRAM (Package on Package)	128 MB	512 MB	4 GB
	Other device specs		
Storage	16 GB	64 GB	Up to 512 GB
Display	480 × 320	960 × 640	1792 × 828
Camera	2M pixels	8M pixels	12M pixels

FPGAs

Field Programmable Gate Arrays (FPGAs) are frequently used in embedded devices. FPGAs allow designers to design a chip from a large number of subsystem building blocks. They are less costly to manufacture than SoCs, because a fabrication mask set and fabrication processing is not needed. System designers can program FPGAs themselves. With the large silicon area that is currently integrated into an FPGA, parallel technology is often employed. One application domain that has been researched intensely since 2000 is computer vision (see Section 4.6 and [HDDR16]). Also, FPGAs as supercomputing accelerators have been explored since the first half of the 2000-2010 decade when microprocessor cores as well as floating-point units could be embedded inside one FPGA. Maxwell was one of the first clusters in which each node contained an FPGA [BBBC07]. Progress has not been rapid but, with the emergence of neural net computing in 2018, Microsoft and Intel launched Project Brainwave in Microsoft's Azure cloud computing fabric. An important consideration in using FPGAs has been the need to use a programming notation that includes FPGA hardware design and layout. Current "C-to-gates" compilation is complex because, in addition to getting the program correct, one must understand which statements will generate hardware and which will execute on already generated hardware [THBC18]. See [Mitt18] for a case study of a parallel application, neural nets, on FPGAs.

ASICs

Finally, one step more complex in IC design and production than FPGAs is ASIC technology. ASICs, like FPGAs, are best focused on embedded devices that can have large production volumes offsetting the added design and implementation cost. However, supercomputer applications of ASICs are also possible. (See the Anton computer for molecular dynamics in Section 4.5 and the DNNP discussion in Section 5.2.)

2.5.6 FROM NETWORKS OF WORKSTATIONS TO CLOUD COMPUTING

Several interesting mechanisms for accessing parallel programming have emerged since the 1980s. When workstations became popular, many inventive computer scientists connected idle workstations on their Ethernet with software that could enable them be used for embarrassingly parallel computing. The Network of Workstations (NOWs) movement arose in the 1990s [AnCP95]. API notations were developed to schedule and communicate between workstations that could participate in a parallel computation. The next step for some was to buy several workstations and put them into the same rack so that higher communication rates were available. Thus, Beowulf clusters emerged in 1994 [SSBD95], as did system software tools to manage a rack full of asynchronously operating workstations, for example OSCAR in 2000. As Internet bandwidth has increased cloud computing and crowd computing have become a new mode of parallel processing. Cloud computing, such as Amazon AWS (2003) and Microsoft Azure (2008), has become a readily available resource for parallel processing in which nearly anyone can construct a temporary cluster rapidly. For embarrassingly parallel applications, crowd computing platforms, for example BOINC (2002) used originally in SETI@home, invites individuals, called loaners or donors, who have somewhat idle personal computers to allow their system to be used to take on a task from someone else's embarrassingly parallel application. In return, the loaners may someday create an application using the platform to become borrowers.

2.6 SUMMARY OF PARALLEL HARDWARE

Throughout the history of parallel processing a number of now familiar design patterns have emerged for making and improving parallel processors. As an approach to summarizing this chapter, it may be worth noting some of the main patterns that fit historically significant machines and trends. Table 2.6 cites nine design patterns. When one machine is an outstanding example of the pattern, it is noted. Otherwise, a historical trend is cited.

There are a few caveats to realize with this table. First, we do not know what the companies that built these systems were really thinking, i.e., they may have had different or multiple patterns in mind. Second, this table is not an exhaustive list of design patterns. Finally, there is no standard way of describing and naming these patterns.

Table 2.6: Some architecture change patterns used in parallel processing

Architecture Change Pattern	Historic Examples	More Recent Examples
Fresh Start	*Goodyear STARAN*	*Fat-tree interconnection*
This pattern tries to start a new company or a new trend by taking advantage of new technology.	STARAN implemented content addressable memory in a dramatically different manor. It had a "flip" network to transpose columns into rows and back.	TMC did not survive even after they migrated from the hypercube to the fat tree architecture. Many of today's clusters are built with fat tree interconnections.
Implement in hardware	*SGI Geometry Engine*	*Microprocessor vector extensions*
This pattern implements a frequent execution path in hardware to avoid overhead.	The SGI Geometry Engine improved on Evans and Sutherland's pipelined graphics rendering process and implemented it in VLSI.	X86 processors have added and extended SIMD vector registers and instructions over many generations of microprocessors.
Transfer HW to ISA	*Cray-1 vector registers*	*GPGPUs*
A different pattern is to move what was in hardware into the instruction set architecture.	The Cray-1 had a faster clock at its time because memory-to-register and register-to-functional unit in two instructions decoupled the long memory-to-memory architecture.	Nvidia, starting with programmable GPU shaders, has successfully listened to the GPGPU community in transitioning shaders into more general-purpose compute engines.
Quantity optimization	*Cache levels and sizes*	*Hybrid parallelism*
In going from one version to the next, architects try to tune the hardware in the face of design trade-offs that must be made.	Nearly every computer designed has had to determine the best cache architecture. Parallel processors especially face issues of level of sharing and coherency.	Quantity optimizations also occur in blending a hybrid cluster node. How many microprocessors and how many GPUs on each node? ([VSBG18], Section 5.7)

Redesign to accelerate	*AP-120B redesigned to the FPS-264*	*Intel Tick-Tock*
Resigning a component or the whole system to take advantage of newer, often faster and denser, technology is often a good choice.	FPS was successful with the AP-120B. Next, FPS-264 was designed with ECL logic rather than with the latest TTL. In addition to increasing from single precision to double precision, the FPS-264 was five times faster.	Intel successfully applied Redesign to Accelerate in alternating steps for several decades. The first step installed a faster/denser fab process, called a Tick step. Then, the next step would redesign the microarchitecture, Tock. Since 2016 they have shifted to a: Process, Architecture, Optimization model.
"Hyperspace Jump"	*Convex Exemplar, Alliant FX/2800*	*Intel attempts Intel i860, attempts Itanium, attempts ARM*
To react to a slowing of sales momentum, companies can make a dramatic architecture change.	When Convex and Alliant saw the killer micro coming, both decided that they had to change instruction set architectures, to take advantage of the microprocessor trend.	Intel has kept the x86 architecture for decades. They have also tried other ISAs: the i860 of 1989, Itanium in 2001, and ARM replaced i860 in 1997. Although Itanium was used in a system that was as high as 2nd on the TOP500, all of these have been discontinued.
Additional features	*Cray T3E*	*Short message redirecting in the switch nodes*
Adding a new, easy to implement feature can sometimes generate more success than the effort.	The T3E was a major redesign from the T3D. Working with DEC, five subsequent upgrades of the DEC Alpha processor were designed with other changes they more than doubled the clock speed.	Cluster switching nodes have become increasingly sophisticated to accelerate frequent message passing patterns.

Co-design	Multiflow VLIW	SoCs, FPGAs, ASICs
All new systems now are designed and validated with both compilation of workloads and full gate-level simulation.	An example that stands out is Multiflow VLIW where the major question was, could compilers really find enough fine-grained parallelism?	Today, designers have a cluster just to simulate a new design all the way from power-on through OS boot to a workload run. This can be especially important when designing custom or semi-custom ICs.
Benchmarks	*Linpack benchmarks*	*Vision, Speech neural net benchmarks*
Benchmarks are usually the kernel from a real workload. Although they provide system comparison points, they also oversimplify.	Linpack100, Linpack1000, and HPLinpack give comparisons between parallel systems. (For system design traces from customer workloads are used most often.)	Vision and speech recognition benchmark have been used in comparing new SIMD instructions, tensor cores on GPUs, and DNNP configurations [RDSK15], Section 4.6.

CHAPTER 3

Programming Notations and Compilers

In 1980, uniprocessor vector supercomputers were the dominant type of parallel machine. As a result, the dominant programming notations were the two most frequently used to program these machines: vector extensions to FORTRAN and conventional, scalar, FORTRAN, translated into vector form by compilers. Today, there is a much wider range of possibilities including the two just mentioned.

To keep the discussion on programming notations to a reasonable size, we focus on the notations most widely used by practitioners today: array notation, automatic vectorization/parallelization, SPMD notations, message passing primitives, shared memory primitives, macrodataflow graphs, and subroutine libraries encapsulating parallelism. Below, a separate section is devoted to the history of each one of these seven notations.

These programming notations can be used separately or in combination. For example, DMM machines with multicore nodes are often programmed using both message passing primitives and shared memory primitives. As a second example, consider autovectorization, which can be applied to single-threaded, shared-memory, distributed memory, or hybrid distributed/shared memory programs.

As is the case with most programming constructs, all the parallel programming notations discussed next trace their origin to ideas and prototypes introduced before 1980.

3.1 ARRAY NOTATION

ILLIAC IV, one of the earliest parallel machines, used SIMD processors with array instructions and therefore it was natural that one of the first notations developed to program these machines was based on array extensions to FORTRAN and other languages.

Array notation was not originally designed to represent parallelism but to facilitate programming by raising the level of abstraction. This is achieved by abstracting implementation details. For example, the C loop

```
for (i=0;i<n;i++)a[i]=b[i]+c[i]
```

would be represented in array notation, assuming all array are of size **n**, simply as **a=b+c**; which neither uses the **for** loop nor the loop index, **i**.

One of the earliest languages incorporating array expressions was Iverson's "A Programming Language" (APL) [Iver62]. APL was designed around powerful operators, especially array operators. The first implementation of APL was released by IBM in the 1960s. An illustration of the importance given at the time to this new language is that IBM supported APL's unusual character set with custom designed keyboards and a Selectric typewriter ball. Today, more than half a century later, array notation is highly popular for scientific computing, especially in dynamic languages such as MATLAB, R, and NumPy. However, today's array notations, like APL, are not primarily used to specify parallelism, but for programmability.

The situation was different for the languages used in the early days of parallel computing. In fact, in the 1970s array notation was introduced to specify parallel operations for pipelined vector processors and SIMD computers. Array extensions to FORTRAN and other languages became the notation of choice to program most of the SIMD machines developed before 2000. Languages containing array operations included: the Burroughs ILLIAC IV FORTRAN [Burr71], the Massachusetts Computer Associates' IVTRAN [Mill73] developed for ILLIAC IV, CFD developed by NASA AMES Research Center also for ILLIAC IV, ICL DAP FORTRAN, Parallel Pascal for the Goodyear MPP [Reev84], MasPar Fortran [MasP92a], and CM Fortran for the Connection Machine [Thin91]. Except for the array assignment statements, the semantics of these languages required results consistent with the execution of one statement at a time in the order specified by the program. This reflected the nature of the target machines in which all parallelism was array parallelism.

Initially, data layout on the SIMD machines was essential but kept to simple rules about how arrays were mapped so that the programmer could understand easily where the compiler would place the data. By the time of CM Fortran, directives were added to allow the programmer to optimize data layout. There was very little local memory on the early SIMDs and one had to optimize its use. (Illiac IV had only 2k 64-bit words per PE. The Goodyear MPP was a 128 by 128 array of 1-bit PEs with 1024 bits each.) By contrast, today's DNNPs can have hundreds of megabytes of data integrated into the array with multiple neural nets all being processed concurrently.

Burroughs ILLIAC IV FORTRAN assignment statements could represent parallelism in the form of element-by-element array operations. Arithmetic operations could operate on complete arrays or subsets of array elements selected using bit vectors, known as control vectors. ICL DAP FORTRAN assignment statements could also operate on array sections selected by bit vectors. Array indices could be slices of multidimensional arrays specified by assigning a specific value to one of the dimensions and leaving the other dimensions empty to represent complete rows or complete columns. The assignment statement in CM Fortran and MasPar Fortran could contain whole arrays or array sections represented using the *<begin>*, *<end>*, *<stride>* triplet notation which is the norm today. Parallel Pascal's array expressions could operate element by element on whole arrays, on array slices, or on a subset of consecutive elements of an array. In IVTRAN, array operations were represented differently using the **do** *<label>* **for all** construct, a precursor of today's Fortran **forall** construct, which can conveniently represent all the array operations that could be represented in the languages just mentioned. (**forall** here is a SIMD operation, not a MIMD parallel loop. So, application developers had to think "keep mostly in lock-step".) An important characteristic is that the **forall** semantics require that the right hand side(s) be evaluated before any assignment takes place.

CDC FORTRAN [Gent82] and Lawrence Radiation Laboratory[1] forTRAN (LRLTRAN) [Zwak75] contained array extensions to represent vector operations for pipelined processors. Like the languages mentioned in the previous paragraph, these two languages could represent element-by-element vector operations in the form of arithmetic expressions. For efficiency of the vector pipeline, sets of elements stored in consecutive memory locations were preferred, but vectors of subscripts as well as bit vectors could be used to represent gather and scatter operations. Although available, it was very important to keep these gather-scatter operations to an absolute minimum on most vector processors because the memory system contained banks that needed to run in a synchronized order. Using them, a supercomputer could end up being about the same as a high-performance serial processor. For example, to be efficiently vectorized car crash applications (Section 4.2) were designed around minimizing their use.

By the late 1970s, the U.S. Fortran Programming Language Standards Technical Committee, known at the time as ANSI X3J3, took notice of the popularity of array-extended version of FORTRAN[2], which includes the ones just mentioned and others such as IBM's VECTRAN [PaWi75], and started considering array extensions for the Fortran standard [Hend79]. This eventually culminated in the Fortran 90 standard; the first standard language to include array notation. Today's Fortran standard is an extension of Fortran 90.

Although its array notation was rarely used to explicitly exploit parallelism, Fortran 90 influenced the design of several of the later array notations for SIMD systems just mentioned including CM Fortran and MasPar Fortran. Fortran 90 was also the base language for High Performance Fortran (HPF), a language to program MIMD machines. HPF used array operations to represent parallelism. By then data layout had become important to specify. The distribution of array elements across nodes of distributed memory MIMD machines was controlled using new declaration statements. HPF was the work of the High-Performance Fortran Forum with more than 60 academic and industrial representatives. The Forum started work in 1991 and delivered the first version of the language in 1993. Although there were initially great expectations and there was vigorous engagement from industry, the project was not successful [KeKZ11] for reasons discussed briefly at the end of this subsection.

---

[1] Today's Lawrence Berkeley Laboratory.
[2] Starting with Fortran 90, only the first letter of the language is capitalized.

Although array notation can conveniently represent SIMD computations and, as HPF demonstrated, MIMD parallelism for SIMD-type computations, array notation is not as popular for parallelism today as it was in the early days of parallel computing. Instead, a number of alternative notations are used for SIMD programming today. For example, to program microprocessor vector extensions such as Intel's AVX and IBMs AltiVec, programmers can use built-in functions or the **simd** directive of OpenMP. Alternately, programmers can rely on the compiler to generate vector instructions from conventional code as discussed in the next section. Other alternatives to array notations are the GPU languages that follow the SPMD model discussed in Section 3.3, notably CUDA and OpenCL.

The fact that array notation is not popular to represent parallelism does not mean that it has disappeared. On the contrary, some of today's most appealing parallel programming notations make use of arrays. One example is TensorFlow [ABCC16], built around array operations that can be implemented in parallel. TensorFlow programs can be mapped onto multiple classes of parallel machines including hybrid machines. For DMM machines, coarrays [DCMC04, Numr11] have gained some traction, but their purpose is to control communication. One or more dimensions of a coarray can be used to read or modify data on non-local memory locations. In this way, conventional assignment statements can be used for communication, avoiding in this way the need for messages. Parallelism in coarrays is expressed in the SPMD paradigm.

Seismic processing, an application discussed in the next section, can use array notation effectively. Some of the seismic processing techniques were first implemented when array notations were first developed, although more so on the Cray than on the SIMD processors. The main reason why array notation is not more widely used to represent parallelism seems to be the semantic gap between the machine independent array notation and the actual code that must be generated for efficient execution. Complex compiler optimizations [KnWL93, KeKZ11] are needed to bridge this gap and achieve the performance and efficiency expected by the programmers of high-end systems. However, today's compiler technology is not sufficiently effective. In the case of HPF, an additional reason for the lack of success was that the language, at least in its initial form, was not capable of representing well some complex parallel algorithms [FoxG18].

3.2   AUTOMATIC PARALLELIZATION FOR CONVENTIONAL LANGUAGES

The use of conventional languages to program parallel machines, relying on compilers to transform sequential code into efficient parallel form, is an appealing idea. It frees the programmer from concerns about the parallel capabilities of the target machine, which as we noted above had to be thought about when using array notations on the SIMD processors of that time. In this sense, using conventional languages raises the level of abstraction over that of explicitly parallel languages. And, when it works, it makes the code portable across machine classes. Compilers that convert conventional code into the vector parallel code of pipelined processors and SIMD machines are called *autovectorizing* compilers. Those that convert conventional code for execution on MIMD machines are called *autoparallelizing* compilers.

Autoparallelization and autovectorization optimizations, as well as the algorithms used by the compilers for instruction level parallelism (ILP), are based on the notion of *dependence* that was described for the first time in 1966 for sequences of assignment statements [Bern66]. The notion of dependence was extended to loops by researchers from the University of Illinois [KuMC72] and applied to vectorization, parallelization, and other powerful loop transformations such as tiling. Work on dependence technology has continued to this day and has led to a formal foundation of the technology [Bane93]. Although as mentioned at the end of this subsection, there is still much room for improvement.

One of the first autovectorizing compilers (and perhaps the first), the Paralyzer, was developed for ILLIAC IV [PrJo75]. However, except for the compiler planned for the Burroughs Scientific Processor, it seems that no other production autovectorizer for SIMD machines was ever developed. The Cray and CDC compilers [Gent84] of the early 1980s implemented *autovectorization* for pipelined vector processors. Array notation could be used to program the CDC machines as well, but for Cray systems, autovectorization was the only way to exploit parallelism from high-level languages [WikCTS]. Autovectorization was considered a success because it allowed new users to try the Cray-1. Cray experts could narrow their questions to a handful about the application to give tentative feedback to "try the vectorizer" or not. Some of the frequent questions were: how many iterations of the loop are there; the array dimensions of the arguments are not specified, can we put in numbers; where are the COMMONs shared; can key subroutines be inlined or are they known to be parallel. A drawback then, just as today, was that experts often had to interpret what the cryptic vectorizer diagnostics meant. At least they could first run the vectorized code to see if the program was faster. With other programming notations, such as array notations, one had to commit to fully recode a part of the application by hand before anything could be tested.

One of the earliest autoparallelizing compilers was implemented for the Alliant multiprocessor in the mid-1980s [TeMS87]. Parallelization of scalar operations emerged in the 1990s along two main directions. The first focuses on ILP [LFKL93, RauR94, SCHL03]. The second, which targets reconfigurable FPGAs [AtSi93], is much less widely used although interest is growing as testified by the recent announcement of Intel's High Level Synthesis tool.

Today, focusing on microprocessors, the most widely used Fortran and C compilers contain autovectorization and autoparallelization capabilities. Autovectorization is currently almost exclusively used to generate the SIMD microprocessor vector extensions (originally called multimedia extensions, MMX, see Section 2.5.5) such as Intel's AVX and IBMs AltiVec. For Pipelined Vector-SMM hybrid machines, the autovectorizing/parallelizing compiler had the task of selecting the best mode of parallelism. For a multiply nested loop, vectorizing the innermost loop and paralleliz-ing one or more outer loops, if possible, was the natural mode. A single loop could be converted into a multiply nested loop by strip-mining and then outer parallelization and inner loop vectorization can be applied. Multicore microprocessor can apply these strategies today. We should mention that autoparallelization, although typically coexisting with vectorization in today's compilers, seems not to be widely used, based on compiler developers' lore, since there are no publicly available statistics on this.

Programmers could enable compiler vectorization that did not pass the dependence tests by adding directives such as **ivdep** and *noalias* or by following simple programming rules. For exam-ple, for vectorization to work, the Paralyzer required that all subscript expressions be of the form $I+C$, $I-C$, or C where I is a loop index and C a loop constant expression. A perennially useful rule is to avoid cross-iteration dependence in the innermost loop. An IBM technical report from 1985 [DuSK85] lists a number of programming rules to enable compiler vectorization.

Astute programmers were likely the main reason why compiler vectorization was effective enough to be widely used. Vectorization was considered so important to Cray users that experts, instead of turning to assembly code, analyzed how they could restructure their code so that Cray Fortran (CFT) would generate the most efficient code [DoEl84]. Dynamic FEA applications, as discussed in the applications section, when properly structured and with the right directives can vectorize most of the code. Dynamic FEA was one of the first commercially successful applications of the Cray architecture. All the elements of each element type can be processed as vectors. Atten-tion has to be paid to vectorizing the gathers and scatters. (In dynamic FEA elements can come into contact or contacts can be broken during the run and that is not known at compile time, so that vectors of indices to element must be used rather than simply vectors of elements.)

Without programmer intervention, compilers are, to this day, often unable to vectorize code even though only simple transformations are typically needed for successful vectorization [MGGW11]. The need for programmer intervention implies a significant increase in the cost of performance tuning. This cost is magnified by the lack of rules to decide what transformations are needed to enable vectorization. For example, consider the case of induction variables. These are variables that assume values that form an arithmetic sequence. Their computation using increments preclude vectorization, but they can often be eliminated by replacing their occurrences with expres-sions of the loop indices. Although vectorizing compilers can often handle induction variables, this is not always the case and, even for simple induction variables, the programmer must in some cases eliminate the induction variable to enable vectorization.

3.3   SPMD PROGRAMMING

In the Single Program Multiple Data (SPMD) programming paradigm, the same code is executed by all processing elements. Although the term was originally introduced to describe an SMM programming paradigm [DGNP88], it is today also used within SIMD and DMM programming models. In the case of SIMD machines, SPMD programs further work synchronously in the sense that all processing elements execute the same instruction at any given time. SPMD is a dominant programming paradigm today. It was used for SIMD and is being used for many SMM applications. And although they are at opposite ends of the parallelism spectrum in terms of task size, it is being used for nearly all GPU and DMM applications.

Programmers of the 1970s found that besides relying on autovectorization as implemented in the ILLIAC IV Paralyzer, or using one of the numerous languages with explicit array notation, SIMD machines could conveniently be programmed directly. The semantics of assembly language programs for parallel SIMD machines is that each instruction is executed by all processing elements (PEs), or a subset of PEs if a mask is used, and each PE operate on different data. The earliest SPMD high-level programming notation with low level SIMD semantics was the language Glypnir developed for ILLIAC IV [LLBR75].

In Glypnir, as in SIMD assembly language, an operation on *PE variables* is executed by each PE using local values. Control statements were generalized so that, for example, in an **if** statement where the condition contains PE variables the **then** part is executed by all PEs for which the condition was *true*, while the other PEs remain idle. If the **if** statement contains an **else** part, it is executed for the complementary set of PEs while the other PEs remain idle. **While** loops whose condition involved PE variables would be executed until the conditional expressions become *false* for all PEs involved. The inefficiency in execution of conditional statements that is introduced when first the **then** part is executed and after the **else** part is executed is called *branch divergence* [HaAb11]. Branch divergence can add considerable inefficiency to SIMD execution.

Glypnir introduced a new class of languages that was later to include the MasPar Programming Language or MPL [MasP92b]. This class of languages evolved to encompass CUDA and OpenCL that are today the dominant notations for programming GPUs. An important extension to Glypnir and MPL is the introduction of shared memory features in CUDA and OpenCL. The main reason why Glypnir-style programming notation superseded the array notation used for the SIMD machines seems to be efficiency. In fact, going back as far as the late 1980s MasPar's MPL Language Manual explicitly tells us that this is the case when it states that MPL was the "most efficient and most flexible" programming language supported by MasPar.

SPMD notation is also widely used today to program distributed memory and shared memory machines. The reason for its popularity is the same: provide better control of the code and thus facilitate achieving high performance. Since in this case the target is a MIMD machine, each node can follow its own path through the code, avoiding in this way the branch divergence penalty of SIMD machines or GPUs.

The common use of the SPMD notation for both SIMD and distributed memory programming seems to indicate an influence of the first over the second although we have not found any acknowledgement of this in the literature.

A GPU SPMD implementation of the molecular dynamics nonbonded computation mentioned above can be found in [SHPS16]. Here, atoms are arranged in tiles to be executed in thread blocks. When computing the force on an atom, if the distance to an atom in an adjacent tile is small enough, there will be a tangible force between them, and that atom is included in the force computation. This distance condition is an example of SPMD branch divergence.

3.4   MESSAGE PASSING PRIMITIVES

Distributed memory machines were not that important in 1980 although, as mentioned at the beginning of Chapter 2 on hardware there were a few *loosely coupled multiprocessors*, as distributed memory machines were known back then. Today, message passing on DMM systems is being used effectively in many applications. For a simple example, sequence alignment applications run efficiently on DMM systems because a genetics database of sequences can be spread across a cluster, e.g., BLAST discussed below in Section 4.4. With the dominance of clusters in the TOP500 illustrated in Section 2.5.1, most supercomputer applications have been structured with message passing, frequently with much more complex techniques than genetics such as domain decomposition used in FEA and molecular dynamics; see Sections 4.2 and 4.5.

The message passing paradigm was developed before 1980. See, for example, [Lori72] on parallel programming with IBM's channel-to-channel adapter and [Hoar78] on communicating sequential processes (CSP). However, the message passing notations of today originated in the early 1980s with the Caltech Cosmic Cube [Seit85],[3] developed by the Caltech Concurrent Computation Program (C3P). The Crystalline Operating System (CrOS), developed for the Cosmic Cube, made message passing operations available to programmers not as a language extension, such as those in CSP, but as a library. This library approach was adopted by practically all widely used message-passing systems developed afterwards. CrOS also introduced to the world of DMM programming collective communication operations such as broadcast and reduction that increased the usability of the notation. The designers of CrOS advocated and used the SPMD paradigm to program the Cosmic Cube [FoxG18].

After the Cosmic Cube, there was a proliferation of message passing APIs. Some developed by vendors such as IBM's EUI [BaKi93, BBBC94] and Intel's NX [Pier94], others by research groups such as PARMACS (developed at the German National Research Centre for Computer

---

[3] The Cosmic Cube was the first Hypercube connected DMM system and it led to production of the iPSC.

Science) [CHHW94] and PVM (developed at Oak Ridge National Laboratory) [Sund90]. An important characteristic introduced by the end of this wave of message passing systems was system independence. Message passing libraries were not integrated into any one OS or any one interconnection network. Thus, they were portable and still pretty effective across different systems. PVM was for a while the de facto machine independent standard for distributed memory programming.

The most popular notation today to program distributed memory machines is MPI (the Message Passing Interface). Development of MPI started in 1992 at a Workshop on Standards for Message Passing and, under the guidance of a committee known as the Message Passing Interface Forum, it created a portable API that incorporated features of the earlier, syntactically incompatible APIs mentioned in the previous paragraph [MPIF93]. This standardization effort resembled the Parallel Computing Forum initiated several years before by Kuck and Associates Inc. (KAI) to standardize shared memory programming notations. See the next section. Today's MPI is much larger than the CrOS library in terms of the number of routines and their complexity, but it retains the CrOS' library reliance on collective communication and SPMD programming style.

A second approach to programming distributed memory machines with message passing primitives is to follow a partitioned global address space (PGAS) model. Among the popular examples of this approach are SHMEM, implemented as a library, initially introduced by Cray for the T3D, and two parallel languages: Coarray Fortran [DCMC04, Numr11] and UPC [ElCh01]. The most recent versions of MPI contain one-sided communication primitives that are needed to implement PGAS notations. PGAS notations have benefits over the first MPI versions, but for efficient execution, they require low latency interconnection networks, for example the Cray Gemini interconnection network; see Section 2.5.4.

3.5   SHARED MEMORY PRIMITIVES

Long before 1980, primitive operations were introduced to control task creation, and synchronization for shared-memory machines. In application terms, the 1980s automatic speech recognition originally experimented with shared memory primitives because the structure of phoneme graphs could not be cast as vectors. (Parallelism in speech recognition is discussed in Section 4.7.) Today, this programming notation is used in many applications and one can use libraries, compiler directives, or language extension implementations.

The most basic programming notations for shared memory parallelism are APIs to control task spawning and synchronization. Mechanisms to represent task parallelism are typically provided by the operating system processes and they have existed since the beginning of electronic computers to support all forms of parallelism including that between I/O and computation. One of the earliest API designs for large task parallelism that is influential to this day is the Fork-Join model whose description was published in 1963 [Conw63]. Also, long before 1981, programming languages such as Algol 60 and PL/1 included tasking primitives. The advent of SMM machines in the early 1980s led to a variety of APIs which supported parallelism for performance including the *macrotasking* functions from Cray [Cray84] that were used to spawn processes for tasks using relatively slow operations. Popular APIs for tasking today include POSIX Threads, which are lightweight compared to operating system processes, and the standard libraries of languages such as Java and C++. There are also tasking APIs for Python although their effectiveness is constrained by its Global Interpreter Lock (GIL). Since not all operating systems contain standard tasking APIs, application developers are forced to use wrapper layers to facilitate portability across machines with different tasking APIs. The specific synchronization primitives their application depended on could also be added to these wrapper layers.

The basic multitasking mechanisms have some limitations as a programming notation. One is that a naive use of the multitasking mechanisms could often result in unstructured programs that are difficult to read, let alone debug. This lack of structure is even greater in programs that reuse threads from a pool of threads. Thread pools are used to avoid the significant operating system overhead of repeated thread creation for each task within the application. For that reason, computer vendors developed collections of directives to represent parallelism. One class of directives (e.g., Cray microtasking directives) annotated loops, typically Fortran loops, as parallel. The idea of a *parallel loop* had earlier been proposed for the Burroughs FMP multiprocessor [LuBa86] under the name of DOALL, but the notation of parallel loops was proposed even earlier [Gosd66].

Starting in 1987, under the leadership of KAI, representatives from the main multiprocessor computer vendors, under the name of the Parallel Computing Forum, undertook the creation of a standard to unify the numerous sets of Fortran directives which included those from Alliant, Convex, Cray, Encore, IBM, and Sequent. The final version of their proposed standard was released in 1991 [PaCF91] and contained not a set of directives, but Fortran language extensions that implemented the constructs that were typically specified using directives before PCF. With the release of the standard, the PCF ended and a committee was organized to develop an ANSI standard based on the PCF proposal. Unfortunately, no standard was produced. As far as we could determine there were two main reasons for this. One was that the PCF founders found the process required to develop a standard within ANSI was too long and cumbersome [Kuck18]. In addition, the ANSI committee was too ambitious and expanded the initial PCF primitives creating what many believed was an unmanageable proposal [Leas18].

After it was clear that the ANSI subcommittee was not converging, KAI restarted the process and assembled a group from companies and research laboratories that put together a proposal for FORTRAN directives and function calls to specify parallelism, including loop parallelism. The primitives included in this proposal were almost the same as those in the PCF document, but with some simplifications. The proposed programming notation, which they called OpenMP, was presented at Supercomputing in 1997 [ChJR07]. The name OpenMP reflects the desire to make the programming notation standard/open access. Under the guidance of the OpenMP Architecture Review Board, the standard has evolved for the last 20 years to include numerous new constructs for C, C++, and Fortran. Instead of using language extensions, OpenMP operations are specified via directives that take the form of **#pragmas** in the case of C and C++ and special comments, in the case of Fortran. Aside from programming with threads directly, OpenMP has become the most popular standard for shared memory parallel programming.

Alternatives to OpenMP for C/C++ are implemented as template libraries in Microsoft's Task Parallel Library (TPL) and in Intel's Threading Building Blocks (TBB), released in 2006 [Rein07]. Because C++ supports templates, TPL and TBB do not require compiler support. In this way, they are simpler to implement and easier to evolve than OpenMP which does require compiler support. Another emerging alternative is the C++17 Standard Template Library, which includes the concept of a parallel iterator. This allows the programmer to apply considerable freedom in creating SMM tasks. Like threads, it is important to observe that APIs, directives, and templates do not check for correctness, which can be a source of problems. These checks can be implemented by applying dependence checking or validating proper synchronization with tools such as Intel Inspector.

Notations for shared memory programming require run-time support in both user mode and by the operating system. For example, task scheduling techniques such as self-scheduling [Smit82], guided self-scheduling [Poly87], and gang scheduling have been an important component in the implementation and use of parallel loops. Also of great importance is the capability of the operating system to switch contexts to a different process if processors become idle during execution of a shared memory program [Boot18, Furt18]. Context switches are used as a last resort in high performance computing since processor idle cost is not as significant as it was for the vector processors of the first wave. (We personally note that they still become important when we introduce a deadlock or busy wait, the bane of shared memory primitives, and start hoping for a context switch to let us get control again.)

Finally, one can compare shared memory primitives to both finer and coarser parallelism models. Today, shared memory primitive programming notations are suitable for multicore processors but are too coarse for manycore processors such as GPUs where a fine-grained SPMD model is appropriate. Threads at this level are referred to as microthreads because the tasks are too fine-grained to schedule individually. Here a block of threads is scheduled concurrently and the threading model guarantees that all threads in the block are alive with registers and cores allocated concurrently. Synchronization is limited in this model to barriers where all threads must check-in to the same barrier before any thread in the block is released. This is in contrast to the threads of the shared memory primitives where synchronization can cause a thread to busy wait, or even be context switched. ([SGMA10] illustrates this model and how it can be translated to shared memory primitives.) An even coarser parallel model such as in MPI that uses processes can easily use multicore today. One can ask for any number of virtual MPI workers on the same multicore, including more workers than cores because a worker will be suspended to allow other workers to make progress.

3.6 DATAFLOW

Although there are no commercial dataflow machines today, the notion of a task graph to control parallelism and schedule computations has been highly influential. At the fine grain, instruction level parallelism controlled by hardware, pervasive today, can be considered an incarnation of dataflow although it predates dataflow and is not explicitly based on a graph. On the other hand, we can legitimately say that macrodataflow, where each node of the graph executes multiple instructions, was partly inspired by dataflow concepts. There have been numerous macrodataflow designs since 1981 and the notation is still of great interest. An early example was the dependence-driven computation design planned for the experimental Cedar multiprocessor, built by the University of Illinois [GaKP81]. Charisma++ [KaHH04] is an example of macrodataflow as a programming model for distributed memory machines. The Open Community Runtime is a recent example of macrodataflow [MCCS16]. This design was intended as an alternative to MPI for exascale computing. Other examples include the dataflow graph mechanisms of TBB, Microsoft's Task Parallel Library (TPL) and TensorFlow. In TensorFlow, the nodes represent array operations and the values flowing along the edges of the task graph are arrays or tensors. Similarly, for applications one does not have far to look. Automatic speech recognition traditionally used phoneme graphs and more recently, deep neural nets models have used TensorFlow. In fact, deep neural net processors, discussed in Section 5.2, rely on macrodataflow notation as their high-level language.

3.7 PARALLEL COMPUTING LIBRARIES

Libraries encapsulating parallelism within the compute intensive functions are of great importance. They facilitate parallel programming in the same way that conventional libraries facilitate conventional programming. If an application can be implemented in terms of subroutines encapsulating parallelism, excellent performance can be obtained without the need for the library user to explicitly handle parallelism or synchronization. Of course, this does not work for all applications, and relying on sequences of basic parallel algorithms does not always lead to efficient computations.

We are not aware of any widely used libraries of computational subroutines designed to encapsulate parallelism before 1980. The version of LINPACK available in 1980 seems to have been automatically vectorized within appropriate subroutines. The Numerical Algorithms Group (NAG) library was "partially vectorized" for the Cray-1 in 1983 [WikNAG]. Also, after 1980, an assembly language implementation of the BLAS for the Cray Y-MP and Cray 2 [SVYM91] was developed and this implicitly meant a vectorized version of LINPACK [DoSt84] since it is implemented on top of BLAS. Even then applications were taking advantage of numerical libraries' parallel capabilities. Static FEA, discussed in the Section 4.2, uses solver kernels and algorithms frequently found in subroutine libraries.

During the last three and a half decades the number of libraries encapsulating parallelism has increased and they collectively now span a wide range of applications, algorithms implementations, and target machines. The libraries sometimes target a single class of machines (shared memory, distributed memory, or GPUs) or include alternate routines for different classes of machines. A few general-purpose examples include:

- IMSL and NAG for shared and distributed memory machines,

- LAPACK for shared memory machines and vector processors (originally LINPACK),

- ScaLAPACK for distributed memory machines,

- NVIDIA GPU-Accelerated libraries,

- Intel's MKL library, which includes implementations for shared and distributed memory machines,

- the Parallel Boost Graph Library for distributed memory machines,

- the C++ parallel Standard Template Library for shared memory machines, and

- TensorFlow for machine learning and neural nets on clusters and GPUs.

There are many application specific libraries that encapsulate parallelism as well.

3.8   PARALLEL PROGRAMMING NOTATIONS AND MACHINE CLASSES

To conclude this section, we summarize in Table 3.1 the parallel programming notations discussed here. The top-5 rows summarize the characteristics of the different programming notations that are the columns of the table. Below that, the relation between machine classes and parallel programming notations is shown. Example languages for each class of notation are shown in each cell indicating that the notation is appropriate for that class of machine.

Table 3.1: Examples of prominent programming notations used by various classes of parallel system. (Parentheses indicate currently little or no current use. "√" indicates some of those above in the column were or are inherited.)

Programming Notation Characteristics

	Array notation	Automatic parallelization	Shared memory primitives	Message passing primitives	Libraries
Format	Compiled language extensions	Conventional languages and compiler directives	Compiler directives or API	API	API
Parallelism control	Transparent	Compiler directives	Parallel regions with workshare constructs	Embedded in SPMD model	Embedded in API
Parallel region	Within a statement	Mainly parallel loops	Identified parallel regions	Whole program	Library functions
Synchronization	Barriers at the end of array statements.	Inserted by compiler	Barriers, and critical regions for reductions	Asynchronous except messages	Hidden in library functions
Data placement	Transparent	Transparent	Caching of data structures	By processor identification	Transparent

Classes of Systems

	Array extensions	Auto-parallelization	Shared memory	Message passing	Libraries
Major examples					
Vector processor (Cray-1, FPS AP-120B, FPS-264, Supertek)	(CDC Fortran, LRLTRAN, Fortran 90)	(CVC)			NAG, LAPACK, EISPACK

Machine class						
Vector processor - SMM	(Cray XMP, YMP, Cray 2, Convex, Alliant)		(Alliant Fortran)	(Burroughs FMP)		✓
Latency hiding processor	(Denelcor, Tera)			HEP Fortran		
VLIW processor	(Multiflow, Cydrome)		(Multiflow Fortran, C)			
SIMD	(DAP, GdyearMPP, ILLIACIV, MasPar, TMI) GPUs	(IVTRAN, DAP Fortran, MasPar, and CM Fortran)	(Paralyzer)			
SMM/UMA	(Sequent, Encore, Celerity) **Multicore**		**Intel Fortran and C, IBM AltiVec**	(PCF) **OpenMP, TBB**		✓
SMM/NUMA	(BBN, Origin, Exemplar, KSR)			KSR Fortran		✓
DMM	(iPSC, nCube, CM-5, iWARP) **Clusters of Multicores with GPU accelerators**	(HPF) *Lisp* **CM Fortran**			(CrOS, EUI, NX, PARMACS, PVM) **MPI, PGAS**	**ScaLAPACK**

CHAPTER 4

Applications

In 1980 there weren't many parallel applications. Those that existed were mostly experimental in research labs. But then, parallel machines were few and they cost millions of dollars. Now there are many parallel applications and we have to select a small representative set to cover in this chapter, see Table 1.1 or 4.1. By 1980 most of these applications had already made a start in parallel processing but the progress was very different. Parallelism in seismic processing started in the 1960s on IBM attached vector processors and was using vector supercomputers on production seismic surveys by 1980. Finite element analysis and computational chemistry were beginning to see their potential [GlHo83, FiRa81]. As early as the 1960s computer vision researchers saw that 2D processor arrays such as Illiac III could work well for image processing [McCo63]. Object recognition, on the other hand, had not seen much opportunity for parallelism because algorithms were still being explored. In 1980, computer graphics experts were experimenting with graphics systems on the Cray-1 and on a DMM system [EwMa78, Fuch77]. By 1981, the E&S CT5 flight simulator came out which had a pipeline of custom VLIW processors controlled by a PDP-11. The whole simulator cost around $20M when a Cray-1 cost $8M. Bioinformatics was researching laboratory equipment and algorithms to support them, a long way from having any interesting results.

Today, with the cost of parallel processors in the $100 range, parallel applications are abundant. And, advances in parallel hardware and programming notations since the 1980s have allowed parallel application developers to achieve many major milestones in their fields. As a result, parallelism impacts millions of people every day. Table 1.1 shows key milestones for these seven parallel applications. Each application is discussed in the subsections below following the temporal order of the milestones. At the end of this chapter, Table 4.2 will summarize the application scalability growth.

What about the engineers and scientists that brought these parallel applications into existence? Most of them are now pre-eminent in their fields. Some looked for ways to use computers to solve problems they could only attack with pencil and paper before that. And, it turned out, they needed a supercomputer in the end. For example, theoretical chemists Walter Kohn and John Anthony Pople were awarded the Nobel Prize in Chemistry in 1998 for inventing ab initio computational chemistry, which is covered in Section 4.5. Others were simply solid engineers with a good idea that they pursued with a passion. For example, John O. Hallquist at LLNL invented the finite element analysis applications NIKE2D, then NIKE3D and DYNA2D. These led to an even better application. In 1976, he invented DNYA3D that we will speak about in Section 4.2. A final example shows that sometimes supercomputing applications have had to win out step-by-step over analog or mechanical technology. Industrial Light and Magic (ILM), with a render farm of over 7500 nodes, was founded in 1975 when George Lucas hired John Dykstra, who did work on *Close Encounters of a Third Kind*. As of 2016 ILM had won 16 Best Visual Effects Oscars and 24 Scientific and Technical Awards from the Academy of Motion Picture Arts and Sciences. See ray tracing in Section 4.3. Thus, for over 40 years, passion to solve real-world problems has inspired many good engineers and scientists to employ parallelism in their applications.

These seven applications have exhibited four orthogonal modes of parallelism growth over the period of four decades.

1. New Parallel Methods (NPM) have been applied. For example, [LTEK20] traces major algorithmic improvements in the maximum-flow problem over 40 years leading to 9 orders of magnitude increase in performance on graphs with 10^{12} nodes.

2. Simply application Data Size Grows (DSG). In some cases this comes about because the number of data points grows, e.g., genetic sequences, or because the precision of the input is increases, e.g., more elements in an FEA model.

3. Additional Scalable Layer (ASL) of parallelism that contribute new functionality. The parallelism in this new layer was typically just repetitions of multiple instances of a problem. This is what is sometimes called "embarrassingly parallel" and more appropriately called "pleasingly parallel" because it is not at all embarrassing to compute and apply more useful computation.

4. Sometimes Time-to-Solution Pressure (TSP) demands best met by using parallel processing have been added.

Historical analysis shows that often an application that benefited from one parallelism mode can grow scalability in the future following the same mode. But, that may not always be the case. For example, a protein has only a certain number of atoms; simulating more atoms is not necessary. Although a large protein can be over 500,000 atoms, a small protein is about 2,000 atoms. But also, in most of these cases, there are more problems to solve, e.g., adding new methods to increase the accuracy of simulation or simply simulating even larger proteins. Advances in a different area of the application field often appear to create future scalability growth in a mode that was not used in the past, e.g., iterating over a set of proteins in a database.

Table 4.1: Applications in Chapter 4 and their major parallelism modes. (In this chapter, the abbreviation for each mode of parallelism will be used to identify the cases. For example, DSG mode will stand for Data Size Growth.)

Application— Major Milestone Achieved	New Parallel Methods (NPM mode)	Data Size Growth (DSG mode)	Additional Scalable Layer (ASL mode)	Time to Solution Pressure (TSP mode)
Seismic Data Processing—3D seismic processing	✓	✓		
Finite Element Analysis—Car crash simulation	✓	✓	✓	
Computer-Generated Images— Full-length animated movie	✓	✓		✓
Bioinformatics— Sequencing the human genome	✓	✓	✓	
Computational Chemistry— Protein folding	✓	✓	✓	
Computer Vision— Autonomous vehicle	✓	✓	✓	✓
Speech Recognition— Continuous recognition with less than 5% error	✓	✓		✓

Before going into the specifics, here are several general points about parallel applications.

- Parallel applications affect ordinary people in two ways. Quite a few parallel applications touch people indirectly. These are usually datacenter applications where experts interpret and apply the results of the computations to benefit many non-experts. For example, chemists must analyze the results of computational chemistry applications to select the most promising compounds for human use. On the other hand, in more recent cases, multicore microprocessors and GPUs in personal devices have allowed parallelism to be used by people directly in daily life, e.g., a weather forecast.

- The seven applications discussed in this chapter have sufficient usage and parallelism to target several different platforms by using several of the programming notations discussed above. The platforms can be simple or hybrid parallel systems. For example, seismic processing, although vectorized in the 1960s, was programmed for the pipe-lined vector supercomputers in the 1980s, a simple parallelism system (see Section

3.2). There they relied on vectorizing compilers. It evolved across many years of version after version evolution onto hybrid parallel systems, adopting multiple programming notations such as: SMM microtasking of Section 3.5, NUMA SHMEM notation of Section 3.4, clusters with MPI of Section 3.4, and more.

- For each application field there is a suite of programs used in various combinations to address a broad range of problems. Each program in the suite might use different algorithms and thus might have different types and amounts of parallelism. Although some of the programs receive little value from parallelism, the most compute intensive components of the suite are often the best parallelized. Only a few applications in each suite can be mentioned specifically below. For a general picture consider two examples (the highly parallel components are in bold):

  ○ Chemical design suite which usually contains:

    - A graphical molecule builder

    - A chemical and protein database interface

    - **Several computational chemistry packages** (discussed in Section 4.5)**:**

        - **Simulated docking tools**

        - **Molecular modelling**

        - **Quantum chemistry analysis**

    - Molecule visualization and property extraction tools

    - **Statistical analysis package for computing chemical properties**

  ○ Computer Aided Engineering usually consists of:

    - A pre-processor to convert computer generated models into a mesh of elements for analysis

    - **Analysis applications which are usually parallel:**

        - **Finite element analysis**

        - **Computational fluid analysis**

        - **Manufacturing analysis**

        - Multi-body dynamics and kinematics

- Post-processing: to visualize the results

- **All of the above are used in an additional layer of parallelism for doing design optimization**

- There has been considerable research into general techniques to improve parallel programs. Consider the three main techniques as follows.

  ○ Tuning for the architecture: For example, today one may wonder why MPP NUMA systems lasted for a decade after the appearance of clusters. Shared memory systems, even with NUMA protocol overhead, were simply easier to program. If one bought a cluster, it could take a couple of years to convert and tune one's applications to DMM message passing. The cost of person-years of engineering was greater than the cost difference between a cluster and a NUMA system. (To make this point to eager DMM parallel processing start-ups, prospective buyers could give the salesperson a set of 10 benchmarks and say, "Bring back your best performance numbers in one month," i.e., not nearly enough time to reprogram an application with message passing. From time-to-time one got lucky and the DMM implementation was easy. More frequently, tuning an SMM application for NUMA could be done within a short period of time. Even today, NUMA architecture is used on a smaller scale, e.g., UltraPath Interconnect.)

  ○ New algorithms: Better algorithms can go as far as making an application a "killer app." To continue the NUMA versus cluster example, implementing domain decomposition in an SMM application could allow it to scale an order of magnitude faster on a NUMA system. And, domain decomposition can make it feasible to add message passing to run even faster on a cluster. Gradually, domain decomposition diffused into many applications and the advantage of NUMAs disappeared.

  ○ New parallel methods: On the other hand, new methods can sometimes launch a whole new market. Here are three examples.

    - Seismic processing: wave migration methods can allow geophysicists to confidently predict whether there is an oil reservoir or not in the survey area. The cost of a well that doesn't produce oil, a "dry hole," is high. The U.S. Energy Information Administration reports that onshore wells typically cost more than $5M. Offshore wells may be more than ten times that cost.

- Finite Element Analysis: the explicit method, as opposed to the implicit method, made engineering cars for crashworthiness possible. Without computer simulation, crashworthiness is experimental: build a car, crash it, rebuild it, …. We note below that a conservative estimate of $2.5M was saved each year by just one manufacturer at the introduction of FEA car crash.

- Computer vision: As noted below, deep neural nets made previous methods obsolete by cutting error rates more than in half.

The figures above are individual instances. When a whole industry applies these methods the multiplier creates large markets. For example, in computer vision, this level of error reduction has made autonomous vehicles feasible. Auto manufacturers call autonomous vehicles the most dramatic change in their industry since the invention of the automobile.

There is a lot of research available about each of these three techniques. But, as just mentioned, the greatest opportunities are in finding new parallel methods. Here are a few pointers to parallel methods surveys: in linear algebra, see [GaPS16]; in deep learning, see [BeHo18]; a final example, the Thirteen Parallel Dwarfs [ABCG06].

- In practice, parallelization of an application is relatively easy compared to the work in preparing the application for production and commercial use. The latter involves a great deal of testing and validating that the application solves necessary real-world problems as well as making sure that it always recovers gracefully from any failure in parallel mode. (Application developers usually make sure the serial version is correct so that they can diff to the parallel version. So, naturally, the serial version almost always has priority for commercial application developers.)

Before discussing these seven applications, here is a short list of other parallel applications with high end-user impact.

- Weather forecasting: many people consult a weather forecasting service before leaving the house. Massive parallelism has been exploited in weather forecasting.

- Climatology: Global warming has such a tremendous economic impact that it touches everyone. Parallelism in climatology has allowed many more climatology scenarios to be simulated and evaluated.

- Computational finance: Financial markets have changed considerably since the 1980s due to computational finance where evaluating risk before a trade is made is essential. Parallelism allows many possible market outcomes to be evaluated too.

- Computer Security: Security analysis with parallel processors helps us to stay one step ahead of fraudulent use that can impact massive numbers of people. And on the other side, the new abundance of commodity parallelism helps perpetrators break security measures faster.

- Oil/Gas Reservoir Simulation: Petroleum exploration is discussed below. Parallelism is also important on the production side.

- Computational Fluid Dynamics: CFD is one of the first applications to employ parallelism. Now, that parallelism has migrated down to most high-end desktops and even laptops, engineers can employ CFD in many quite ordinary design problems.

- Data Analytics: Data analytics is used in many fields more than bioinformatic laboratory analysis mentioned below. From its roots in statistics and data mining, as data size has scaled-up, data analytic methods have employed more and more parallelism.

- Defense and National Labs: Computational analysis of national goals such as natural resource management, productivity, and defense has been primary driver of parallel processing since the computer was invented. Cray-1 serial number 1 was installed at Los Alamos National Laboratory in 1976, while Livermore National Laboratory received the CDC STAR-100. Summit is located at Oak Ridge National Laboratory. And, there are many examples of national labs supercomputers in between.

Still more parallel applications with high impact to society could be listed. See, for example, [HPCA18, FCCS92, ACCI11].

Here are two general notes to keep in mind about this chapter.

- In this book, the term "application" usually refers to an application field. Experts in a field tend to call each program in the field "an application." That latter sense of the term may be used in this chapter when discussing different specific types of programs in an application field.

- There are generally two types of references provided below. First, a technical reference that shows one way in which parallelism works in the application. Second, a historical reference as early as possible to a parallel version on some parallel architecture.

4.1   SEISMIC DATA PROCESSING

The goal of seismic analysis is to allow geophysics to understand the geology of the surveyed area ever more clearly and ever deeper into the earth. The processing power that is applied to seismic surveys has increased tremendously since 1980, in fact, since the first computers. This need for processing power is mainly due to ever-larger seismic surveys, starting from 1D surveys and progressing to 3D surveys today (DSG mode).

Seismic processing of these surveys consists of a sequence of methods: first signal enhancements, such as deconvolution, is applied; continuing with corrections for the geometry of the survey, e.g., normal move out; and, most significant in accuracy and computational complexity, wave equation methods. Wave equation methods are governed by the transmission and reflection of seismic signals. Of the three types of methods, seismic migration is the newest parallel method (abbreviated NPM as discussed above). Traditionally, parallelism has been applied by processing many shots in parallel. (A shot, or trace, is the data that corresponds to one sound source and microphone pair. Hence, a shot was often recorded as one input record.) With this technique, parallelism in most parallel architectures, vector machines, SMM machines, DMM machines, SIMD devices, and GPUs can be applied. And often, they can be applied in hybrid combination. For example, in 1989 on a Cray X-MP the Cray macrotasking as well as vectorization were used in 1989 [ReEd89]. At that time, 3 × 5 mile 3D survey took up to 6000 reels of 6250 bpi tape and that most of the seismic methods just mentioned, except migration, were processed while loading the next tapes. Progress has continued constantly. For example, in 2012, prestack depth migration and reverse time migration was performed on a cluster with GPU accelerated nodes 10–20 times faster than a cluster with CPU nodes alone [Zhan12].

In 1970, it was shown that geophysicists could visualize the earth's structure in a more natural way with 3D surveys, one could say in a stereo view. This was a golden moment in seismic data processing. This insight propelled further progress:

"The geophysical firm CGG officially entered the Big Data market in 1971 with its first 3D seismic acquisition survey for the oil and gas industry

...

"A land seismic survey conducted in 2005 had 400,000 sensors per square kilometer; by 2009, that number had reached 36 million. From 2005 to 2009, the average volume of data gathered on an eight-hour shift grew from 100 GB to more than 2 terabytes.

...

"Marine seismic surveys are more complex, with a vessel towing a streamer a kilometer wide and a 10K trace distance in length. CGG uses multiple vessels to create wide and full azimuth views to open the second eye and see in stereo below the sea floor surface. These streamers represent the largest moving infrastructure on earth, and can be seen from space as they move and collect data.

...

"The company has 40 computing centers worldwide; one of its larger computing centers is within the [size] range of major processing centers such as Sequoia [located at Lawrence Livermore National Lab], with more than 100 petabytes/day of active data."

Comments by Hovey Cox, Senior VP of Marketing and Strategy, CGG [Boma15].

It is interesting to note that large-scale machines such as the Cray XC30 provide several advanced parallel programming notations, e.g., Fortran co-arrays, UPC, and Cray SHMEM notations, as well as MPI, to allow geophysicists to program the most recent seismic analysis techniques and numerical methods [Bran17]. However, we do not know whether these advanced programming notations are finally in use in production geophysical application suites.

Since the origins of parallel processing, seismic processing was always the first to look for more parallel processing. As a historical note, the "S" in the TI-ASC processor name originally stood for "Seismic" and was changed to "Texas Instruments—Advanced Scientific Computer" when they found other applications might also be interested. The future of seismic processing appears to demand even more computational power. Much more data can be integrated to provide ever-clearer views below the surface of the earth. More valuable insights can come from integrating data from sources such as: more flexible sensors, multiple surveys, and satellite remote sensing indicating data size growth (DSG mode) and potential additional scalable layers (ASL mode). Further, new or more complex algorithms can be applied such as: more sophisticated migration, principal component analysis, convolutional neural nets, and self-organizing maps (NPM mode) [Kris15]. Driving this growth is an extension of what geological structures are being searched for. Until recently, one looked for specific types of structure likely to "trap" oil as it migrated through the sub-surface. Now there is increased interest in identifying interesting "continuous" structures where oil is likely to be generated although it may migrate elsewhere.

Although demanding more computational power, one could say that the computational problem of seismic processing is mostly solved. New methods are being developed to help analysts avoid missing opportunities, but the basic methods are stable. New problems are also well defined and in the works. One could look at the U.S. Geological Survey world assessment of oil and gas resources to see not only where conventional seismic studies have been done (40–50% of the world) but where unconventional studies have been done (only 10–20% of the world). Conventional analyses look for particular geologic structures known to trap oil or gas. Unconventional studies usually look for continuous oil and gas accumulation, where new methods are being developed to identify "sweet spots." (Undoubtedly, society will not pursue these resources rapidly because of global warming.)

4.2   FINITE ELEMENT ANALYSIS

Finite Element Analysis (FEA) applications have an interesting and mixed record in parallelism. The focus of this section is on NASTRAN and DYNA3D that are used largely for different types of mechanical structural analysis. NASTRAN (NAsa STRucture ANalysis) was first released in 1968. Shortly after MacNeal-Schwendler Corporation released a commercial version, MSC/NAS-TRAN. NASTRAN uses an implicit solver method, which is often used for linear static and vibration modal analyses. It was first vectorized for the Cray-1. In [GlHo83], it was noted that unless the analysis was large (for those times), the performance on the Cray-1 was not much better than that of a high-performance scalar processor of the same period. Although the element assembly phase is also parallelizable using an SMM tasking notation, it was not considered effective. SMM machines were not widely used in those days and the solver took considerable time. Similarly, gathering all the elements of each type into vectors could have been effective parallelism, as DYNA3D showed several years after. Nonetheless, NASTRAN showed improvement in the solver phase on vector architectures using a parallel library. This is due to the dominance of vector-vector and vector-matrix operations, BLAS Level 1 and 2 [LHKK79]. Since then, more parallelizable numerical methods are being adopted in these solvers, such as preconditioned conjugate gradient techniques (NPM mode). This type of FEA method has advanced to support all architectures from DMM systems to GPUs [AnTD16]. Multicore and GPU parallelism in personal computers has enabled small engineering companies to profit from parallel FEA analysis, for example see [FiZe16] in which ANSYS, another popular application for implicit FEA is used.

DYNA3D uses an explicit solver FEA method for solving dynamic, i.e., time varying, and nonlinear analysis problems. It was implemented in 1976 and shortly after vectorized on the Cray-1. In 1988, the commercial version, LS-DYNA, was launched by Livermore Software Technology Company. Vector parallelism was part of the software architecture from the beginning [Bens07, GiJo89]. The Cray Microtasking notation was also applied as soon as it was available on the Cray-XMP. As discussed in Section 3.5, Cray microtasking enabled the efficient execution of parallel loops by reusing a thread when it had finished the previous task for the next task and in this way significantly reduced the number of expensive task creation operations. Although Cray-1 style parallelism was not widely available at the time, it was the target architecture of choice because dynamic FEA analysis with sufficient number of elements for valid analyses could rarely be performed on slower machines. A major milestone occurred when automotive structural engineers applied nonlinear dynamic analysis in the design of safer cars during the 1980s and 1990s. Several safe car features credited with saving many lives every year have been designed on parallel computers. For example, crush zones, door beams, and air bags. A key requirement of this type of problem is that the finite element model evolves over the duration of the simulation with new contacts and separations between elements expected to occur all the time. Every major automaker has several hybrid clusters continuously simulating car crashes. Hybrids can be used efficiently since the support of parallel architectures in LS-DYNA has grown to include SMM machines, DMM machines, and GPUs. It was estimated conservatively in 1999 that roughly 30% of car crashes could be simulated with LS-DYNA instead of crashing a real prototype. Based on the number of real prototypes crashed, an estimated $2.5M was being saved each year by just one manufacturer [Bens07]. The number of elements in complex structures such as automobiles has increased through the years. Around the turn of the century a car model contained typically 250K elements. Now, more than 1M element models are common (DSG mode).

Over the past 40 years, more powerful computers have supported the demand for higher quality engineering in the automotive industry and elsewhere with all types of FEA applications: static, dynamic, modal, linear, and nonlinear [Gins99]. And, design optimization, which adds an additional scalable layer (ASL mode) to FEA analysis, although optimization algorithms themselves are only moderately parallel, the different types of mechanical analysis methods used in the optimization are parallel. Optimization is now available with many FEA applications.

In the near future, parallelism in FEA codes is likely to become more sophisticated for two reasons. First, topics such as optimization and multiphysics, both of which dramatically increase the computing required (ASL, NPM modes), are continuing to develop [KMWG13]. In multiphysics, new methods are being reported by integrating CAE applications, and also multiphysics applications, for example, COMSOL. All these cases, lead to new low cost parallel application demands. Second, the democratization of FEA, i.e., use for many new, different types of product, which has occurred through the availability of FEA applications, has led to a schism between generic FEA applications versus fit-for-purpose FEA [Wass15]. Democratization implies that with respect to parallelism the largest portion of FEA use may continue to move from large DMM servers to SMM and hybrid GPU clients. Multiphysics and optimization certainly require more computation but there are computational dependences in dataflow that make it hard to predict that computation growth will lead to great parallelism.

FEA has more solved problems than unsolved problems. An example of the newer challenging types of problems just mentioned in multiphysics and integrated CAE is designing motors for electric airplanes. This benefits from solving electrical, fluid dynamic and mechanical equations in the same simulation. This is an important problem area because for short flights, 100 miles or less, conventional planes are the most inefficient. They consume fuel costing around $400 while the same electric plane flight is anticipated to cost $12 for the electricity. Start-up companies, such as those investigating sustainable technologies, are tackling problems like this. In the past, they would have been hard pressed to use CAE methods. Multiphysics methods required large computations. With the democratization of CAE due to multicore/GPU workstations, start-ups can make faster progress and save precious investment capital. However, large companies with big servers may not see great scale-ups.

4.3   COMPUTER-GENERATED IMAGERY

The parallelism in Computer-Generated Imagery (CGI) touches almost everyone daily. CGI includes computer graphics, computer games, computer animation, and related applications. In this field, historically, there was a bifurcation between real-time polygon rendering for interactive use and photorealistic rendering, which requires high computational intensity to achieve the appropriate visual realism. Although today both types of CGI applications can be executed on GPUs, originally GPUs were designed for real-time polygon rendering on personal systems while SMM machines and later DMM machines were used for photorealistic rendering.

As noted above in Section 2.4.4, parallelism and graphics have had a long partnership. Parallelism in real-time rendering started in the early 1960s. Rudimentary parallelism was in the form of graphic displays connected to an additional processor to offload graphic processing from the mainframe. Before 1980, parallelism within the processor had been applied to CGI. [EwMa78] discusses a real-time graphic system where the Cray-1 built a display list, which was transferred to a PDP-11/70 and then an Evans and Sutherland Picture System 2 (TSP mode). Also before 1980, [Fuch77] studied parallelism in graphics on a distributed memory, multiple microprocessor system. In the early 1980s, the first special purpose graphics accelerators were attached to workstations, e.g., Clark's Geometry Engine which had a 12-stage rendering pipeline. See [Croc97] for the history of techniques in parallel real-time graphics. GPU history started in the 1980s with processors for polygon rendering in arcade games. Into the 1990s, 3D graphic accelerators were common in game consoles and on PC graphics boards.

In contrast with polygon rendering, photorealistic rendering changes the design criteria from rendering speed to rendering quality. Photorealistic rendering in computer-generated animation has perhaps impacted an even broader public than polygon rendering. A few landmark computer-generated sequences in conventional films were:

- 1973: *Westworld*,

- 1982: Both the *Genesis Effect* sequence in *Star Trek II: The Wrath of Kahn* and the movie *Tron*, and

- 1984: *The Last Starfighter* which had 27 minutes of CGI produced on a Cray X-MP.

A major challenge to film makers has been integrating CGI with traditional camera-film sequences, an early glimpse of augmented reality. Landmarks here are films such as:

- 1991: *Terminator 2: Judgment Day* and

- 1993: *Jurassic Park*.

In the mid 1990s, a major milestone was achieved, the first fully computer-generated feature length film (DSG mode):

- 1995: *Toy Story*.

See [WikTCA].

The fundamental ray tracing algorithm goes back to 1968 [Appe68]. Ray tracing became the main algorithm for photorealistic rendering around 1980. Ray tracing is considered embarrassingly parallel in that the shading of all pixels can in principle be computed in parallel. However, the limitations of real parallel processing make it harder to get perfect speedup. For pipelined architectures, ray tracing was vectorized on the Cray-1 and Cyber 205 [ArKi89]. In the late 1980s, Ray tracing was parallelized on MPPs [Dela88] and on DMM machines [Muus87]. Finally, ray tracing was parallelized for GPUs around 2005 [FoSu05]. With respect to current computer resources used for ray tracing, in 2017 a Pixar animation render farm of 80 renderBOXXs containing 960 cores was typical of the industry. For production rendering [Sala17] notes:

"... *Render farms often exceed 10,000 cores.* ...".

Hybrid parallel systems distribute frames at DMM level and ray tracing a frame at the SMM level (NPM mode). To support higher performance on modern SMM hardware with its multi-level cache, geometric data structures for ray-polygon intersection are used. Today, real-time ray tracing with acceptable quality has been close to becoming a reality on GPUs for a few years [MSGS18]. However, the rapidly changing geometric data structures that occur in real-time gaming and visualization, can cause data structure construction for shared access to become the performance bottleneck. Further improvements in data structure construction have been achieved [HLJH09] so that real-time ray tracing is appearing in games [Nvid18].

The future of parallelism in CGI is bright. Content creation, as in the past, has a strong agenda in datacenters. For example, Amazon Web Services includes the Zync Rendering Service. At the same time, content creation for HDTV is striving for real-time bandwidth [Fraz17] creating time to solution pressure (TSP mode) on the parallel architecture. Looking forward, Virtual Reality and Augmented Reality will probably have near term impact on client devices as well as in the cloud. Even further, although in large part also image processing, 3D telepresence and tele-immersion are promising and still in their infancy [CGNG15, KuYB16] (NPM, DSG modes).

Although the usage of computer graphics is bright, CGI is mostly a solved problem, especially polygon rending. However, real-time ray tracing is becoming fast enough so that it can replace polygon rendering on client devices. One can add augmented/virtual reality and telepresence to that. For example, 3D reconstruction, which goes back at least to 1913, can finally be performed on commodity devices in these new application areas. 3D reconstruction with the goals of achieving real time presence, with few cameras, and low transmission bandwidth has already triggered many proposed solutions.

4.4   BIOINFORMATICS

One of the milestones in science and health of the last 40 years was sequencing the human genome. The oft-cited *Science* article on sequencing the human genome [Vent01] discusses their need to parallelize the key routines for gene assembly. At that time, the late 1990s, The Institute for Genomic Research ran on 10 four-processor SMM machines plus a 16-processor NUMA system within their 440-processor Compaq Alpha TruCluster. (One should note that there was and is parallelism in the laboratory processing of DNA material as well as in the computational analysis of DNA information. In the laboratory, a large number of DNA sequencing machines work in parallel, e.g., Sanger Sequencers [HeCh16]. This produces genetic sequences called reads for DNA computational analysis.)

Today, bioinformatics has evolved to address many questions of great social importance, often with the use of parallel computers. DNA computational analysis, genomics, touched on in this section would have for example, as an objective to provide genetic sequences causing a new disease in an organism. Then, the genetic sequences are assembled and used to search a large database for closely matching genes or their associated proteins to enlarge the human disease information network. In 2014, the European Bioinformatics Institute stored 40 petabytes of data on genes, proteins, and smaller molecules. They ran on a 17K core cluster. And, the amount of data was doubling every year [KHAH16] (DSG mode). (Keep in mind that one individual's genome is around 200 GB.)

For this application field, a key algorithm is sequence alignment, a.k.a. gene matching. For example, sequence alignment is used repeatedly to assemble reads into longer sequences. And then it is used again to compare the longer sequence to sequences in a genomic database. Today, several sequence alignment algorithms are used. See [Stad79] for one of the earliest descriptions of how this matching works. There is usually ample parallelism opportunity no matter which alignment algorithm or parallel architecture is used because the genetic material being matched can be partitioned to match against a database of other independent pieces, which can also be partitioned for execution of many sequence alignments in parallel. In the foundations of bioinformatics in the 1990s, parallelizing on SMM systems was found to be relatively easy using this partitioning technique, as mentioned above regarding sequencing the human genome. Later, other parallel architectures were targeted. For example, the Smith-Waterman algorithm has been parallelized for microprocessor vector extensions like MMX in Intel Processors [RoSe00]. This used C++ with inline assembly code notation.

BLAST is probably the most popular alignment algorithm [AGMM90]. Although it is not the most accurate alignment algorithm, it is fast with reasonable accuracy. Today, very large genetic databases such as GenBank are partitioned and stored across a large DMM cluster. For another level of parallelism, often a website interface allows many researchers to query the database simultaneously (ASL mode). [DaCF03] discusses mpiBLAST and [ZLZM11] compares mpiBLAST running on an in-house cluster to mpiBLAST running on an Amazon AWS cluster. BLAST has been parallelized on most architectures including GPUs [VoSa11]. New parallelization techniques for BLAST are still appearing, for example [MLBV17].

There are many different genomic applications and more appearing frequently. Quite a few of them can be posed as algorithms using sequence alignment to build relationships among pieces of genetic material (genomics) or between proteins (proteomics). Most of these scale well, but not completely. For a survey of frequent parallelism motifs in these applications, see [YBAA20]. Below we will mention a very computationally complex application of sequence matching, computational phylogenetics.

Bioinformatics has several very rich fields of microbiology research in which many applications and algorithms are data-rich and parallel. It is an experimental science involving many aspects of microbiology in which big data and data analytics play a major role [Adam15]. Big data analytic tools such as MongoDB, Hadoop, Spark Streaming, and Mahout are used to name a few (NPM mode). Macrodataflow notation tools such as GraphLab may be used to control the flow of large lab dataset computation on a cluster. For example, [KHAH16] discusses, among other bioinformatics data analytics techniques, association rule mining of DNA expression studies using DNA microarrays. (A microarray contains thousands of comparison dots of gene sequences, for example from a healthy subject, and from an experimental sample, perhaps a cancer sample.) Data mining is used to find associations, a frequently occurring pattern of genes expressed among the thousands of dots paired in the two samples. What makes this a "big data" problem is that a subject would have a series of DNA samples taken at several points in time. And, there may be one series of samples for each of the 100s of subjects in a clinical study. [WSSA18] illustrates some of the bioinformatics techniques applied to current medical research.

This field is far from being solved. DNA sequencing has grown from decoding the human genome to mapping the genome of all living things. This map can be organized as deriving the evolution of all living things. A frequent problem in the new area of computational phylogenetics is to find a good or optimal phylogenetic tree linking 10s, 100s, or 1,000s of taxa. (A taxon is a population of an organism seen by taxonomists to form a unit. Taxa is the plural form.) A very relevant example is to model the evolution of pathogenic viruses such as HiV, Ebola, or coronavirus to understand how infections may be treated, e.g., each of these viruses is one taxa evolved among many less potent related virus taxa. These problems are much more computationally complex than the problems discussed above. Not only is data growth rapid in this area, but new methods are also sought. Nearly perfect speed-ups on large clusters are reported for some phylogenetic search methods, but speed-ups due to better methods, "only" in the range of 200,000-fold is still extremely parallel.

4.5   COMPUTATIONAL CHEMISTRY

Computational Chemistry (CC) has always been seen as an analysis methodology that could grow to be of the same caliber as other the types of chemistry analyses, *in vivo* (in real life), and *in vitro* (in the laboratory). A new term was coined for computational chemistry, "*in silico*." Like Bioinformatics, CC consists of many different applications. Let's consider three of these application types.

The first of the three is Molecular Dynamics (MD), which models a molecule in terms of the forces between atoms in the molecule and in adjacent spaces. The first research paper using the MD method on a molecule of 864 atoms appeared in 1964 [Rahm64]. This simulation ran on a CDC3600. (A computer designed by Seymour Cray.) Today, with this type of application one can study very large molecules such as whole proteins.

MD applications simulate many time steps of the loop: compute the forces on each atom due to surrounding atoms, integrate these forces, move the atoms in the direction of the total force, and repeat. There are several types of atomic force that must be calculated, but the most compute- and communication-intensive is the electrostatic force between atoms. Fortunately, the larger the molecule, the more abundant the parallelism is because there are more atoms to simulate, and the forces between atoms decrease with distance allowing mostly simultaneous simulation of the atoms that are distant [PZKK02, PeSc17]. Applying a heuristic *cut-off distance* beyond which electrostatic force computation can change from atom-to-atom computation (communication intense) to a long-distance force computation allows a spatial meshing, a partitioning of the molecule, so that the force created by all the atoms in each partition can aggregate their effect on distant atoms (reducing long range communication). All the atoms in a partition can be computed in parallel followed by a computation of the long-range effects.

From 1980 until now, one could easily increase the number of atoms modeled whenever a larger parallel processor was available. (The smallest human protein is about 2,000 atoms. The median is about 7,200 atoms. But the largest, Titin providing muscle elasticity, is about 642,000 atoms. To which one adds at least surrounding chemical solvent models of interest, e.g., a model for water. But, more often one simulates multiple molecules interacting at active sites. Simulations between 100,000 and 10 million atoms are typical.) That is, one could increase the speedup by weak scaling, which purportedly "beats" Amdahl's Law (DSG mode). Parallelism in molecular mechanics started with vector processing [FiRa81] and has continued on all types of parallel architecture including DMM systems with GPU accelerated nodes [SHPS16] and with ASIC nodes. The basic parallelism strategy has evolved as the number of cores has increased. On vector supercomputers, explicit lists of atoms within cut-off distance were used. Then, all the atoms were in the same shared memory and atoms on the list could be accessed randomly (although slower than for strided accesses). Today, on clusters, atoms in mesh structures should be used with additional computational methods, such as Particle Mesh Ewald, which uses 3D FFTs to compute long distance forces (NPM mode). In most cases, a parallel programming notation appropriate for the architecture has been used because the key kernels are well known and not too difficult to recode.

A major milestone of MD has been to simulate a protein chain as it folds into the complex 3D shape, which defines how it reacts with other molecules. MD is the primary tool in this research. The Anton supercomputer [Shaw09] set a record in 2010 for simulating 1.5 milliseconds of a protein folding in a run taking over 100 days of computation. Timesteps in MD are measured in femtoseconds. This run simulated more than 100 times longer simulated time than had previously been simulated [Ledf10]. Anton, built at D. E. Shaw Research, is a special purpose 512-node ASIC cluster, connected in a 3D torus. Each ASIC contains two subsystems: the first has 32 deeply pipelined arithmetic modules; the second has 10 Tensilica cores with 8 SIMD cores each. Anton improves the performance of a general-purpose cluster running MD by: tuning the interconnection network and protocol specifically for MD communication (Anton uses a 3D torus configuration); using special purpose cores in of the subsystems, each tuned for specific types of MD force calculation; and finally, using a lot of computation and communication and relatively little memory so that the Anton memory subsystem can be streamlined.

The second application type for CC is based on quantum mechanical simulation. Ab initio computation was developed before computers shortly after the Schrödinger Equation in the late 1920s. Some of the major types of these applications are ab initio, density functional theory, and semi-empirical. By simulating the principles of quantum chemistry with the Schrödinger Equation, ab initio applications are much more computationally dense modeling electron wave functions with basis sets in atoms in smaller molecules, e.g., inorganic chemistry.

Ab initio computation is highly parallel and the parallelism can be implemented with shared memory parallel loop and/or message passing notations [MAKD17]. Until large parallel processors appeared, molecules of only a few atoms could be studied. MD is used for modeling much larger molecules spread over a greater distance, e.g., organic chemistry. In effect, MD data structures are sparse, while ab initio uses dense matrices. A major commercial ab initio code, Gaussian, was benchmarked on a 512-node Cray T3E in 1994 reporting speedups up to 33 on 64 processors [SCOF98]. The resulting speedup depended on: the molecule size, the number of basis functions, and the analysis mode.

For a third type of CC application, a major commercial application that has emerged in the last 30 years is drug design. Drug design brings together several CC applications to carry out *virtual screening*. A combination of lightweight applications is used to evaluate the docking potential of two molecules, viz. their likeliness to react. For example, a dataset of synthesizable compounds can be screened against a database of human proteins to estimate the potential reactivity of each compound. If the potential reactivity is high, the compound can be analyzed in much more detail with other CC applications, such as those above, for an estimation of positive and negative effects before they are taken to the "wet" lab. In the 1990s, databases were 10s of thousands of compounds and the lightest weight applications of the times could be used for screening on SMM systems (ASL). Now, screening is usually carried out on large DMM systems [YuMa17]. Databases may be 10s of millions of compounds, for example ZINC15 has 230 million compounds, so that a few seconds at most should be spent to evaluate each docking and even lighter weight applications are used. In a study of light docking applications, one was selected that took a few seconds of elapsed time more than the others on a single core yet produced better reactivity estimates. It was parallelized in SMM mode to achieve a 7× speedup on an 8-way multicore node and it replaced the use of several of the poorer quality docking applications [TrOl10].

One view of the future of computational chemistry is to propose that at its heart it is using a computer to simulate the composition and reaction of all matter. Which is to say, a vast and parallel CC computation can be envisioned [AsLR18] (ASL mode). However, it is interesting to note that the computers used need mostly lots of processors and relatively little memory. The actual data input is small, a protein is usually much less than a million atoms, the internals of the computation that follows generates large data output demands, e.g., ab initio computations can be 4–7 loops deep and MD analysis of behavior in the femtosecond range drives huge computations. There are also research advances in drug design. It is estimated that the synthesizable chemical space is 10^{30} to 10^{60} molecules [ENSL18] (DSG mode). Research is ongoing into new methods to tackle this huge chemical space in a structured and parallel manner. For example, [PoIT18] studies reinforcement learning methods (NPM mode).

Like bioinformatics, this field is far from computationally solved on the drug design side. Mapping, organizing, the effects of so many molecules is just starting. To take the human out of the loop, it is being viewed as a machine learning opportunity. The same is true for proteins. Humans have envisioned the behavior of proteins by finding and defining many different conformations of the parts of a protein. How these conformational parts form and interact with other molecules is doubly complex.

4.6 COMPUTER VISION

Computer Vision (CV) consists of several major application areas. One of these, image processing, was one of the first applications of parallel processors. Going back to 1958, there were studies of parallel architectures based on 2D mesh interconnection that would execute many pixel-based image processing algorithms efficiently when a rectangular set of pixels was assigned to each processing element [Unger58]. The ILLIAC III [McCo63], a 2D cellular architecture of 36 × 36 processors, was built and programmed, although the image processing then was rudimentary compared to today.

Image processing is an application where the first vector processors, e.g., Crays, were not very cost efficient. Their long word length made it awkward and inefficient to unpack, operate, and repack 8–24-bit pixels into Cray's 60-bit words. (Given the cost of these machines, very high efficiency was expected.) However, image processing covers many applications and attached array processors such as the FPS AP-120B had more manageable physical size and reduced cost so that they were a practical solution for the more complex image processing problems, which may combine signal processing with image processing, for example, tomography, radar, sonar, and satellite temporal data applications [McRK85, KaCo81] (DSG, ASL).

To move on to another computer vision application area, *object recognition* is perhaps the most intensely studied application in CV. Many distinctly different algorithms have been developed since 1980 (NPM mode). Object recognition is often thought of in three stages with the first stage being *image processing* to clean and clarify the image for the benefit of the specific subsequent stages. Middle-stage algorithms, such as segmentation or classification, are mostly as appropriate for SIMD processing as image processing. However, the last stage of object recognition may be executed more efficiently on SMM machines because geometric data structures are used to efficiently represent object relationships and movement in the case of video processing.

A now classic object recognition algorithm is Scale-Invariant Feature Transformation (SIFT) [Lowe04]. SIFT can be used with current systems for video object recognition. That is to say video implies that the goal is real-time or near real-time performance (TSP mode). Hence, like many seminal algorithms in a variety of applications, it has been parallelized on several different architectures such as multicore microprocessors [ZCZX08] and GPUs [FaRo15] using the parallel programming notations of choice for these architectures.

There have been recent milestones in object recognition by using deep neural nets (NPM mode). For example, since 2010 the ImageNet Challenge has posed an object recognition benchmark. In 2012 there was a dramatic breakthrough in which classification error rate dropped from 25% to around 15% when convolutional neural nets (CNNs) were used. CNNs composed of several stages, i.e., Deep Neural Nets (DNNs), with each net being much larger than had been used in prior decades, could be trained using many more images. This marked a transition from algorithm-centric CV, such as SIFT, to data-centric CV. AlexNet, for example, was trained on two Nvidia GTX580s in 5–6 days [KrSH12]. By 2015, GoogLeNet reduced the error rate to 6.7%. It also trained on a few GPUs in about a week [SLJS15]. By 2017, the winning team had a 2.3% error rate [ChLY18]. The DNNs in this case were trained using eight servers or less, with eight NVIDIA TITAN X per server. (Note that ImageNet Challenges are a moving target, changing roughly every year [RDSK15]. There are several different challenges every release, e.g., ImageNet 2017 among other tests includes localization, which can be important in applications such as image understanding in the next paragraph. Note also, there are several metrics for success, e.g., precision, how repeatable the result is, as well as classification error rate. ImageNet is now hosted by www.kaggle.com, which contains many other data science benchmarks.)

A major milestone in *image understanding* is happening now. (Image understanding is more comprehensive than object recognition. It involves some degree of knowledge representation between recognized objects and some degree of inference capability on that knowledge. In the case of autonomous vehicles, the goal involves decision processing, such as "Stop Now, right and left lane blocked", etc.) Some industry experts say that autonomous vehicles may be the biggest change in the automotive industry since the car was invented. An onboard platform for autonomous driving uses parallel processing at several stages.

1. Perception tasks: Fusing and processing data from multiple sources, ultrasonic sensors, radar, LiDAR (Light raDAR), and multiple cameras, while simultaneously performing image processing algorithms. This stage is perhaps best performed on digital signal processors (DSPs).

2. Recognition tasks: Includes object classification, localization, and object tracking. Perhaps best performed on GPUs.

3. Decision processing: This stage includes path prediction, path planning, and obstacle avoidance. These may perhaps be best done on multicore microprocessors.

The best overall system architecture is not clear at this point. Since DNNs were mentioned above, we note that the NVIDIA DRIVE system is designed to compute deep learning methods in the central stage of these automotive applications. However, because power consumption is critical in vehicles, FPGAs (e.g., Intel's Cyclone V) and ASICs (MobilEye EyeQ5) may also be competitive [LTZG17, LZHS18]. It should be noted that for automotive, as well as for image-based Internet of things devices, system optimization with hybrid parallel architectures is indicated.

CV has a robust future. Few if any parallel application areas combine processing high bandwidth real-time data streams (as from one or multiple sources) with high computation requirements (as in the image processing ➔ object recognition ➔ image understanding) into small portable formats. In addition, there are still several research challenges; [Mali19] recently pointed out four near-term challenges for computer vision: few shot learning, learning with little supervision, unifying learning with geometric reasoning, and perception and control. (TSP, DSG, ASL modes). With respect to platforms, as with CGI both edge device and cloud parallelism are indicated. See [DeAr15] for some recent examples of CV applications and methods (NPM mode).

Consisting of two parts, image processing and object recognition, CV is partly a solved problem and partly far from solved problem, respectively. Object recognition in autonomous vehicles is far from solved. Onboard sensors will produce around 1 GB of data every second may require 100 TOPs of computing from multiple SoCs. For example, an Nvidia Xavier designed for this problem, available in 2020, contains: a multicore CPU, a GPU such as Nvidia Turing, a vision accelerator, a deep learning accelerator, and an optical flow engine. On the datacenter level, the problem of improving object recognition to assure safe autonomous operation, using current DNN methods, is said to require recording and analyzing thousands of operating years of onboard camera data, a huge parallel computation process.

4.7   AUTOMATIC SPEECH RECOGNITION

Compared to other applications discussed here, automatic speech recognition (ASR) seems the weakest candidate for parallelism on today's high-speed hardware. However, it is useful to include ASR to show that parallel processing is often useful for improving quality of results. With applications that provide indirect results through domain experts such as seismic processing, bioinformatics, etc., it is hard for end users to see the benefit of parallel processing. However, if one says to their virtual assistant "take me home" and "telephone" is understood instead the need for more processing or a better algorithm is clear. It seems like ASR should not be computationally intense; the input data is a serial digitization at the relatively slow rate of human speech. But there are several complicating factors: the need for speaker independent recognition; the difficulty of recognizing non-native speakers; and the very noisy real world environment. Finally, in many cases one can trade-off accuracy for speed, but ASR exemplifies that in command and control achieving both simultaneously is required. ASR is at a turning point; today's server clusters have improved quality to the point where the general public is accepting ASR. Google reports that over 20% of web queries from mobile devices will be spoken in 2020.

Training recognizers is also an interesting subject for parallelism. Although originally recognizers were phoneme graphs constructed by hand, today recognizers are more accurate and have a larger vocabulary when built on servers using the greatly increased audio corpus available on the Internet.

ASR was a goal of researchers for several years before 1980. In 1952, Bell Labs Audrey could recognize digits spoken one by one. In 1962, IBM's Shoebox was demonstrated at the Seattle World's Fair. In 1971, DARPA funded computer-based speech understanding research. It took quite a few years and several competing fundamentally different methods for limited initial goals to be reached. Four methods were explored between 1973 and 1980: stochastic, logic-based, natural language understanding, and discourse models [JuMa09]. Between 1983 and 1993 probabilistic methods emerged. The Sphinx-II system [AHHH92] scored the highest in the DARPA evaluation in 1992. It was the first system to do speaker independent, large vocabulary, continuous speech recognition with 95% accuracy. Without first having a recognizer with high accuracy at any recognition rate, it made little sense to speedup researched approaches with the expensive parallel platforms of the day. Sphinx-II's success in 1992 was past the vector machines' peak years in the 1980s. Killer microprocessor-based multiprocessors and clusters were just appearing in 1992 and parallelism in ASR could be studied on small laboratory-based systems [TBRT02]. Recently, GPUs have also been applied to ASR [HDYD12].

We should underscore that ASR, particularly the recognition engine, compared to feature/ word extraction, has been a challenge for parallelism due to the irregular graphs present in Hidden Markov Models (HMM) or Weighted Finite State Transducers (WFST). The main challenges have been avoiding redundant computation, avoiding memory conflicts due to irregular data access, and achieving fast work queue insertion and extraction [ChGK11, YCYG09]. On personal devices SIMD instructions and onboard GPUs can decrease recognition time on some tests from 1.6 sec to 200 msec and 66 msec, respectively [HDYD12].

Today, in fact from 2000 onward, deep machine learning approaches to ASR have become dominant due to the accuracy improvements achieved with Recurrent Neural Nets (RNNs) when trained with the much larger corpus of online voice/text data (NPM, DSG modes). A large corpus could only be analyzed with abundant cluster hardware supporting large amounts of memory. By 2015 Deep Speech 2 [AAAB15] trained on a GPU cluster slightly outperformed human word error rate on *Wall Street Journal* articles and audio books. Word2vec is also a frequently cited ASR application [MSCC13]. It should be kept in mind that through the early 2010s, the trained speech recognizers required more memory than normal personal devices supported. This perhaps explains why embedded device ASR has had limited acceptance. Now, high-quality ASR products are cloud-based. Also note that, on these servers batching requests for parallel transcription has been used in keeping average latency low with good efficiency [AAAB15] (ASL mode).

With voice-based user interfaces reaching broad adoption, long-term goals for ASR are pushing beyond speech recognition toward higher levels of artificial intelligence with cognitive and dialog capabilities [Stan16]. (2024 and 2030, respectively, have been cited as goals.) (NPM mode). Although it is quite difficult to say how we will get there, the goals set in [Stan16] are based on current AI research trends and computational resources reaching teraflop scale on a smartphone priced product around 2020. The study goes on to assess that computers will go beyond impacting individuals to impacting several industries, transportation, medical, education, each with their own unique vocabulary (DSG mode).

ASR is closing in on "problem solved." ASR requires the least parallel processing of the applications studied here. It has benefited from parallelism growth by shadowing increases in computing power to attack problems such as limited vocabulary and accents. So that the next larger problem, person-machine dialogs has come into focus. Here, most current commercial systems are task-oriented. That is, the goal of each dialog application is specifically limited to help solve a small fixed set of tasks, e.g., arranging travel or a doctor's appointment. These systems are mostly human intensive frame based, i.e., ontology-based systems using tools such as VoiceXML. A major problem is that, to date, automated learning systems where parallelism might be more beneficial can too easily learn unintended domains of dialog. Without having a way to detect when any of the many inappropriate domains has been learned, it has been best to stick with a frame-based approach. Although natural language processing has been researched since the beginning of artificial intelligence, it's clear that new software methods are needed to automate achieving "machine cognition." For our purposes, one hopes these new methods would be parallelizable.

4.8   APPLICATION SUMMARY

There are many parallel applications today. Forty years ago, there were hardly any and even fewer that made a difference in people's lives. Seven parallel applications were reviewed here that directly or indirectly impact millions of people. Figure 4.1a–c shows the performance growth over time patterns demonstrated by the applications reviewed above. These are general patterns that don't convey precise data; they provide an image that fits the historical trends. The dashed lines in each chart correspond to the pattern for an application's growth over the 40 years studied. The light solid lines are from Figure 2.1 and repeat the performance growth lines of Cray supercomputers, top of the TOP500 clusters, and Intel microprocessors for reference. If an application growth pattern line is above the performance of the hardware line, it indicates that the application could have used even larger systems than were available at the time. If the application pattern line falls below the performance of some type of parallel processor, it indicates that multiple application runs could be run simultaneously on that hardware as a workload. Taken together these patterns indicate that parallel applications' performance has increased at the pace of parallel processors for over the forty years and in the majority of cases these applications are likely to grow as fast as future performance increases in the hardware.

Table 4.2 summarizes the growth pattern found in each application areas in terms of the methods that were discussed in the chapter. It refers the patterns in the different parts of Figure 4.1a–c. The table also lists the application methods that were outlined in each section above. A specific application using each method is shown in the next column. (Note that each application area uses more methods than those described here.)

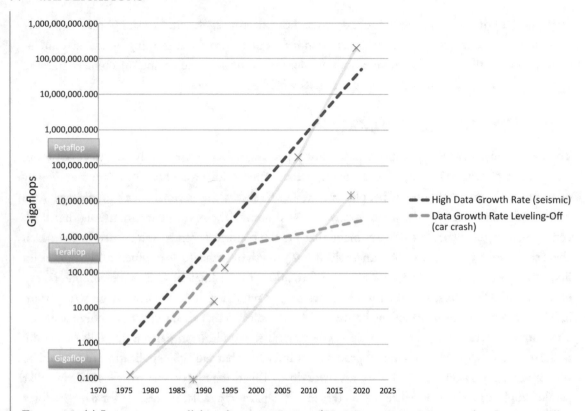

Figure 4.1: (a) Long-term parallel application patterns. (*High Data Growth Rate* applications typically can use all parallel performance increases that occur over time. *Data Growth Leveling Off* applications typically slow their need for adopting new hardware performance increases after hardware performance eventually meets their needs.)

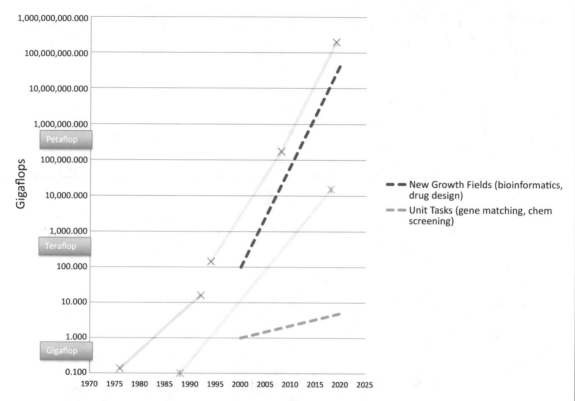

Figure 4.1: (b) Application patterns for new fields. (*New Growth Field* applications typically grow their demand for parallelism faster than hardware performance increases. Being new, and applying parallelism for the first time, they quickly incorporate parallel hardware already available. *Unit Task* type applications launch many unit tasks in parallel, in ASL mode. Also, over time, the unit tasks' size tend to increase slowly as more features are added.)

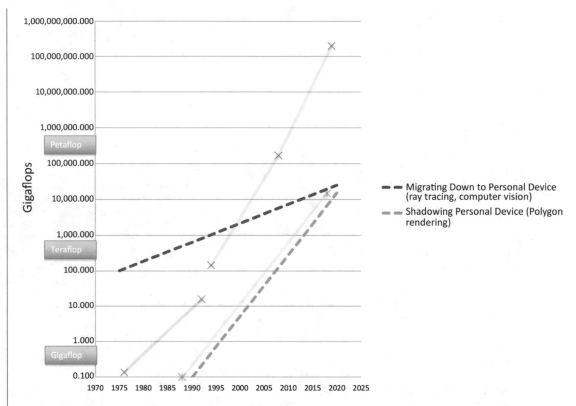

Figure 4.1: (c) Client targeted parallel application patterns. (*Migrating Down To Client* applications typically increase their performance slowly. So that, over time, they can move from the datacenter to personal or embeddded devices. *Shadowing Client* type applications are already on personal or embedded device. They can take advantage of parallelism advances in client hardware.)

Figure 4.1: Application performance growth patterns.

Table 4.2: Parallelization of applications discussed in this section

Application Field	Methods Reviewed	Example Application	Parallelism Growth Pattern
Seismic Data Processing	Digital signal processing	Deconvolution	*High Data Growth Rate* pattern of Figure 4.1(a). Seismic data processing has grown steadily over the period from 1980–2020. It has always been able to use the largest parallel processors available. See the comments of Hovey Cox in Section 4.1.
	Geometry-based processing	Normal move out	
	Migration	Kirchhoff migration	
Finite Element Analysis	Implicit linear solvers	Nastran, ANSYS	*Data Growth Leveling Off* pattern of Figure 4.1(a). FEA applications like DYNA3D grew steadily as FEA models for cars and other structures grew in the 1980s and 1990s. See [Bens07, GiJo89]. The field is still growing as more products from smaller businesses install small clusters and workstations with multicore and GPUs.
	Explicit dynamic solvers	DYNA3D	
Computer-Generated Imagery	Polygon rendering	OpenGL	CGI has two growth patterns. Polygon rendering has shadowed the performance growth of GPUs, the *Shadowing Client Performance*, Figure 4.1(c). GPU performance is approaching the level at which Ray-tracing can be performed in real time, a *Migrating Down to Client* pattern shown in Figure 4.1(c).
	Ray-tracing	BRL-CAD	
Bioinformatics	Sequence alignment	BLAST	In the *New Growth Field* pattern of Figure 4.1(b) the rapid growth of this application is reflected in the growth of the large genetic databases. Each database entry corresponds to a unit task to be performed, e.g., sequence matching. This is an example of the *Unit Task* pattern of Figure 4.1(b).

Computational Chemistry	Molecular dynamics	Anton	This field follows a couple of growth patterns. Molecular dynamics and ab initio computations follow the *High Data Growth Rate* pattern. Drug design is the New Growth Field and *Unit Task* patterns.
	Ab Initio	Gaussian	
	Drug design	Click2Drug [SwIB13]	
Computer Vision	Image processing	StreamHPC	Image processing has two growth patterns *Data Growth Leveling Off* pattern. In traditional areas. On the other hand, high-level consumer cameras follow the *Migrating Down to Client* pattern. Object recognition follows the *New Growth Field* pattern in training neural nets phase. Inferencing on the vehicle, follows the *Unit Task* pattern.
	Object recognition	OpenCV	
	Object recognition	GoogleNet	
Automatic Speech Recognition	Probabilistic model	Sphinx-II	As embedded hardware improves, ASR quality can be improved, a *Shadowing Client Performance* pattern. Datacenter-based ASR follows the *New Growth Field* pattern as verbal grow and as context-based models improve performance.
	Speech recognition	Deep Speech 2	

CHAPTER 5

Parallel Hardware Today and Tomorrow

In this chapter, the major classes of today's parallel systems are compared to historical parallel systems. Then, some research on where the field is headed is cited. The first section asks where are the big parallel processing companies after 40 years. The second takes a look at a new wave of innovation, Deep Neural Net Processors(DNNPs). The reader can contemplate how these DNNPs will fare as businesses compared to parallel processing companies of the past after reading Section 5.1. Sections 5.3–5.6 look at recent progress in the ingredients of the third wave. Sections 5.3 reviews the last 20 years of progress in clusters and looks at exascale clusters on the horizon. Section 5.4 reviews microprocessors since multicore appeared, a multicore that pushes the limit today, and the recent progress in ARM multicores. The architectural concepts behind GPUs compared to their implementation is presented in Section 5.5. Section 5.6 looks at where MCM and IC technology can go, a subject central to parallelism's progress. Finally, we look at where parallel system architecture is going in the near term. We tried to write these recent events from the point of view of: first, the lessons learned in parallel processing in the period from 1980–2020; and second, where parallel system architecture is going.

5.1 SUPERCOMPUTER COMPANIES VS. COMMODITY COMPANIES

As we said at the beginning of the book, parallel system decisions sometimes depend on business decisions. Here, we touch on parallel processing companies as businesses. None of the vector supercomputer companies, not even Cray, survived to be the long-term market leader in parallel processing. And, we don't see a new parallel processing company on the horizon to change that. Start-up parallel processor companies successfully defined and dominated several niches, then they disappeared. In the end though, parallel processing is now solidly established in several major computer and information system manufacturers, IBM, Intel, HP Enterprise, Fujitsu, and others. Compare this record to other computer industry sectors where a dominant company pioneered and survived, e.g., IBM in mainframes, Intel in microprocessors, or Google on the Internet.

The turning point in supercomputing for Cray Research, Inc. (CRI), the largest supercomputer company of the 1980s, was when SGI bought Cray (1996). SGI did quite well with parallelism at first. Founded in 1981, a parallel rendering pipeline for high-performance workstation graphics was a key differentiator from other workstations. In the early 1990s it launched the Challenge series of SMM crossbar switch servers, growing to $3.7B at its peak in 1997. Then, after several years of decline, SGI sold Cray to Tera Computer Company, 2000, and finally, SGI went bankrupt in 2009.

If one allows that the supercomputer bubble of the 1980s was bound to collapse, one should ask if the commodity parallelism companies strongly or fully invested in popularizing parallel processing have done and will do better. We argue the commodity companies with historically strong parallelism strategies have done better over the last 20 years. Table 5.1(a) compares two commodity parallelism companies, Intel and Nvidia, to Cray, Inc, at 5-year intervals from 2000–2019. (One can follow a historical thread from CRI to Cray, Inc. In late 2019, Cray, Inc. was acquired by Hewlett Packard Enterprise.) The last 20 years show that the two commodity product companies have done well. They both have had some unprofitable years also. Cray Inc. also did well in this period growing by 4x on the strength of their high-speed, HPC-oriented, cluster interconnection networks, see Section 2.5.4. Cray, Inc. before they were acquired was comparable in size to CRI 30 years ago when the Cray X-MP was launched.

We don't know why supercomputer companies have had a difficult time surviving. Finally, because their clients are mostly tech savvy and they can change to a new parallelism technology more rapidly than a computer system companies that supplies them. This may be true when the new technology provides more speedup and also when the technology reduces cost, e.g., the birth of clusters. Perhaps, when computer companies have an installed base of loyal customers, they are reluctant to abandon or appear to abandon their current technology. Perhaps the commodity companies have a customer base that would like to buy something easier to use more than something faster. Perhaps high-performance computing is at best an industry sector. And, industry leaders eventually acquire industry sector companies. For example, Intel may explore supercomputing as a technology source for their more profitable commodity microprocessors. Wave one and two parallel hardware companies were not the only computer companies that did not survive. For example, the period from 1980–2020 also saw the birth, growth, and death/acquisition of workstation companies. None of them successfully transitioned to desktop or laptop systems. In conclusion, with parallelism established in several commodity corporations with diverse product bases, downturns in parallel processing as a whole seem less likely.

5.2   HARDWARE INNOVATION—DEEP NEURAL NET PROCESSORS

Over the last 40 years innovations have been frequent. In wave two, recall that the peak parallelism innovation years were from 1982–1984 (Section 2.4.3). Section 2.4.6 reviewed when the wave two start-ups transitioned to microprocessors, the early 1990s. Also note that in the early 2000 decade,

DMM clusters arrived (Section 2.5.1). The next major change, the switch to multicore micropro-cessors, 2006, has not been pointedly innovative. On the other hand, a little noticed innovation at that time, making GPU shaders user programmable, launched the GPGPU and accelerated cluster era.

A new innovative trend starting around 2010 was a wave of start-up companies based on Deep Neural Net Processors (DNNP). (They are sometimes called AI Chips.) Table 5.1(b) com-pares several of these companies. (Note that this table is based on fragmentary information available at the time of publication about each DNNP. The information was collected from: posts on their website, disclosed to technology review sites, or at the Hot Chips Symposium.) Some historical comparisons to these DNNPs are listed here.

- DNNPs are not general-purpose parallel processors the way most previous parallel systems were. They are tuned to perform DNN learning and inferencing by execut-ing a limited set of linear algebra computations. With this specialization, citing one comparison point [JYPP17], they are 15–30× faster than multicores and GPUs and 70–200× more power efficient.

- The architectures of the DNNPs apply several of the concepts of massively parallel processors of the second wave.

  - Short precision PEs, between 8bit integer and 32-bit floating point. Compared to: the ICL DAP—1 bit, later 8 bits; the MasPar—4 bits; or the TMC CM-2—1 bit with 32-bit floating point accelerators.

  - 2D array configuration, e.g., the Cerebras Wafer Scale Engine and the Mythic Intelligence Processing Unit. Compared to: the Goodyear MPP, the Illiac IV, and others.

  - Systolic operation, e.g., the Groq Tensor Streaming Processor is systolic 1D bidirectional. Compared to the CMU/GE/Intel Warp 1D bidirectional systolic processor.

  - VLIW, e.g., the Groq Tensor Streaming Processor and both the Habana Goya and Gaudi. Compared to the Multiflow or Cydrome VLIW processors.

  - Memory distributed into the array, e.g., the Graphcore Intelligence Processing Unit. Similar to node memory in the massively parallel processors and clusters.

- Compared to the early MPP systems there are two major improvements. First, the amount of memory distributed across cores. For example, the Graphcore IPU has 256 KB of static RAM per core for a total of 300 MB per 1,200 cores. Second, because all the cores are on the same chip, the bandwidth of the interconnection networks is much higher, e.g., 7.7 TB/sec cross-section bandwidth on the Graphcore IPU.

- To gain performance density over more general-purpose processors, such as GPUs, DNNPs use shorter numeric data types: 32-bit floating point down to 8-bit integer, often with capability for other types in between. One can see precursors in the MPP SIMD systems, which had to compose larger data types from 1-bit PEs. For example, the Goodyear MPP performed 32-bit floating-point products "only" 3.4 times slower than 24-bit integer products, which were 1.8 times slower than 16-bit integer products.

- DNNPs are used in two-system configurations, which frequently use the vendor's same DNNP chip. One configuration for training neural nets, which uses larger data types, more memory, and larger bandwidth IO interfaces. These systems can be used as nodes of a clustered DNNP configuration. The other configuration, on smaller single board systems, is for deploying the trained networks, a.k.a. an inferencing configuration, which can be either a single chip for embedded systems or a cluster for datacenters. The ICL DAP team produced a deployment version in the Mini-DAP and the Mil-DAP.

- DNNPs frequently compare their performance to GPUs. When executing DNN applications, GPUs use SMT to hide memory latency until the DNN coefficients are in L2 cache. On the other hand, each DNNP has a unique approach to hide and limit memory latency. Principally, they provide memory to store DNN coefficients locally. Some latency variation is still present and some DNNPs also use SMT. For example, the Graphcore IPU provides six-way SMT. On the other hand, the Groq TSP, because it is systolic, is "fully deterministic", meaning, in principle, the number of clock ticks for processing a compiled DNN can be counted and does not change for any input stream.

- DNNPs companies attempt to reduce the software they have to develop and to shorten the time for their technology to diffuse into the market by supporting standard deep learning software tools, such as TensorFlow, and by providing only the necessary machine-specific kernel libraries. Mini-supercomputers in the second wave took the same strategy by supporting only standard Fortran and C via autoparallelization and by tuning only the low-level, popular numerical libraries. A strength of DNNP technology is that DNN programs are relatively small. (In contrast to the small DNN programs, the training datasets are large and they must be marshalled through many DNNP training runs.) The applications of the 1980s were larger and each application had to be vectorized and sometimes parallelized specific to the machine.

The following points are not comparable to historical parallel processors.

- The system software tools are unique for each DNNP. They schedule the DNN macrodataflow graph on the DNNP's resources. A performance monitor must be used during experiments to understand the efficiency of the system.

- DNNPs are fabricated with processes that are about two generations (14–24 nm) behind those used for GPUs and microprocessors (7 and 10 nm). DNNPs still outperform GPUs and microprocessors. [ThSp18] point out the economic advantages of this strategy.

- Each DNNP company tries to present a differentiation to win particular segments of the growing AI market. Differentiation can be with marketing, partnerships, price, or other methods. The following are a few technical differentiations of DNNPs.

  ◦ Cerebras implemented a wafer scale DNNP. The chief benefit is that inter IC delay is greatly reduced. The risks are in keeping the yield high by using programmable routing around a defective die and in cooling a whole wafer uniformly. The latter because temperature differences across the wafer can cause intermittent faults.

  ◦ Mythic uses an analog circuit for matrix multiply. The neural net coefficients are programmed into flash transistors that act as variable resistors of 256 values, providing 8-bit accuracy. This technology makes matrix multiplies both fast and low power.

If one were to take the lessons of the supercomputer start-ups of the 1980s too literally, one would say few if any of the DNNP companies would survive. One of the DNNP start-ups, has already been acquired by a larger company; Habana was acquired by Intel. Another, Wave Computing, recently filed a Chapter 11 to seek shelter from its creditors. As with the second wave, start-ups acquisition or merger is more likely than bankruptcy and the knowledge gained by engineers in DNNP start-ups will benefit the whole industry. (The experience of one of the authors in a second wave start-up that failed was still positive.)

On the other hand, the breakthroughs and applications of DNNs have already been greater than the AI expert system boom of the late 1970s and early 1980s. And most of the DNNPs demonstrate the four key advantages that special purpose processors should demonstrate over more general-purpose processors to be successful [HePa19]: abundant parallelism, high locality, numeric precision adapted to the application, and domain specific languages for rapid diffusion (see Section 5.7).

There are also evolutionary paths for DNNPs. For example, some of these systems may expand into less special-purpose linear algebra appliances to be deployed in new market segments.

Although today DNNPs perform linear algebra computation with single or half precision computation in both training and inference stages, given that DNNPs can migrate to denser fabrication processes, double precision multiply-accumulators are not too far away for them. Once DNN macrodataflow notations mature and diffuse as did GPGPU tools, the next generation DNNPs may be more general-purpose. [LTEK20] points out that specialization of a system is regularly followed by generalizations such as this. Above, we noted programmable GPU shaders as an example of generalization.

A final note, several large enterprises have created their own DNNPs, for example, Google's Tensor Processing Unit [JYPP17], Alibaba's Hanguang 800, Microsoft Project Brainwave, Amazon Web Services Inferentia, and Intel's Stratix FPGAs.

5.3   CLUSTER LESSONS—LEADING TO AN EXAFLOP

After almost 30 years of growth and evolution, where are clusters now and what is next? Table 5.1(c) show the configuration of the top cluster on the TOP500 list at 5-year intervals from 2005–2020. Both the peak performance (Rpeak) and Linpack speed (Rmax) have increased 1,000 fold in 15 years. The memory capacity reported on the node and in the processor cache has increased only 20 fold, surprisingly little compared to the computational speed-up. The network bandwidth of the node has hardly increased at all in the last 10 years. One concludes that supercomputer applications have become more sophisticated in managing memory and communication resources through teamwork. [VSBG18]

Making a large step toward an exaflop, in June 2020, the Summit cluster at 149 petaflops was eclipsed by Fugaku at 415 petaflops, as shown in the first row of Table 5.1(c). Fugaku is located at the RIKEN Center for Computational Science in Japan. (As an aside, clusters are now used in a variety of non-floating-point applications. For example, the Summit cluster had already achieved an exaop performance in 2018. This was on a genomics application [Hine18].)

We are looking forward to the next generation of cluster systems where the race to have an exascale cluster operational is in progress. For example, in 2021 or 2022, the Aurora cluster at Argonne National Lab is expected to be operational after having been delayed a couple of times. There are several echoes of parallel processing evolution anticipated.

- On the SMM side, growing from ASCI White's four processors, each node will have two 30-core Xeon processors.

- The GPU side will have four Intel Xe GPUs connected with an all-to-all switch sharing an integrated 10 PB memory. Similar to early vector pipelined supercomputers, kernels will operate in memory-to-memory streaming mode. However, the kernels will be much more complex than the simple vector instructions of the 1980s.

- Multiple programming notations used on previous clusters will be integrated in the oneAPI development environment: array, autoparallelization, SPMD, and message passing.

- The interconnect will be Cray Slingshot in the dragonfly 3-hop topology. Switches in the fabric will have 64 ports to run at 12.8 Tb/s with adaptive routing software. Historically, Cray has been evolving their interconnection systems since 1992.

There are several other exascale clusters in progress. Less is known about these projects, but to some extent they will be similar to Aurora plus the technology growth likely to occur every year. Here are some examples we found [Andr20].

- The Frontier cluster expects to reach 1.5 exaflops and will consist of more than 100 Cray Shasta cabinets with AMD EPYC microprocessors and Radeon GPUs in a 1–4 ratio. Networking will also be Cray Slingshot dragonfly.

- The El Capitan cluster has been announced. It will have the same capabilities and ingredients as Frontier. It is expected to be in production by 2023.

- Three clusters in China are hoping to reach the exaflop mark before the U.S. systems. They will be built with Chinese components.

- The Fugaku cluster mentioned above is already close to a half an exaflop and is still evolving. It is based on ARM architecture by Fujitsu and RIKEN and uses 7nm Fin-FET technology.

What is important about the direction of clusters? First, they rely on data growth to continue scaling. Note that in Table 5.1(c) Nmax, the size of the linear system used in the benchmark run, has increased by a factor of 12 from 2005–2020. Second, the capacity ratios of: cores, memory, interconnect, and IO must be scaled proportionally to maintain a high fraction of achieved to peak performance. For example, [VSBG18] computes and compares these ratios for Summit, Sierra, and Titan. Then, this paper comments on goals for exascale clusters and applications. Finally, to make cluster applications programmable, software capabilities must continue to improve in the whole stack from interconnect routing, to critical paths in the OS, to programming tools (see Section 2.5.2). For example, Tianhe-2 was criticized as achieving high peak performance at the cost of being hard to program, and Tianhe-2 is not the only example. In 2014, Chi Xuebin, deputy director of the Computer Network and Information Centre said, "Some users would need years … to write the necessary code."

Elsewhere in this chapter, other recent technologies that can support continued scale-up of clusters are discussed: going to the next generation of conventional CMOS devices; 3D MCMs in the node and in the switch; DNN methods, perhaps DNNPs; and programming notations will broaden to facilitate exaflop programming.

5.4   COMMODITY MICROPROCESSORS STILL EVOLVING

Multicore microprocessors keep increasing the level of parallelism, even with the reported slowing of Moore's Law. For example, today's most parallel x86 microprocessor is the AMD Ryzen Threadripper 3990X with 64 cores. This processor has 24 times the performance on the PassMark benchmark as the first quad-core microprocessor in 2008, Intel Core2 Extreme X9770. Note that this Intel processor ran at 3.2 GHz with up to 4.3 GHz in Turbo mode, compared to Threadripper at 2.9 GHZ.

Here are some parallel processing highlights of recent multicore microprocessors.

- Hyper-threading, introduced in 2002, is a form of SMT. Usually all x86 processors run two threads on the same processor now. Thus, the Threadripper supports up to 128 concurrent threads before needing to do context switching on compute-bound workloads.

- Microprocessors started echoing vector processors in 1997 with the introduction of MMX and "3DNow!" vector instructions. Since then five generations of SSE and several generations of AVX up to AVX-512 have followed (see Table 2.3). Threadripper supports AVX2.

- SMM shared cache design has been refined many times on microprocessors. Now, there are three levels of cache with L3 being shared on-chip cache. In addition, cache management protocols have become more sophisticated. For example, Intel Smart-Cache implements L3 shared cache, implements a prefetching policy, and allows cache flushing in the background.

- It should be noted that the Threadripper is an MCM of eight chiplets of eight cores plus another chiplet for IO mounted on a substrate connected by AMD Infinity Fabric. (The first Intel quad core of 2008 was also an MCM.)

This Threadripper is extreme for several reasons. First, recalling that the clock has not increased in recent years, in comparison to the quad core, some increase in performance is due to package cooling improvements allowing more cores to run simultaneously. The 2008 processor was run at 140W TDP (Thermal Design Power) and today's processor is run at 280W. Microprocessors with such high TDP are intended for high-end deskside and desktop systems with active cooling devices. Second, although parallel programming notations will continue to diffuse into more consumer applications, today, whether consumer or professional, only applications that are highly scalable and very compute bound can use that many cores. For example, content creation or high-end engineering workstation applications. A third caveat is that this Theadripper does not have an integrated GPU, even though the additional cores do help graphics performance. Overall, two socket desktop systems, either cpu-cpu or cpu-gpu, having an aggregate greater memory and IO capacity, may cost less, and perform better.

One should also note that microprocessors for servers and non-consumer systems, AMD EPYC and Intel Xeon product lines, are configured differently with: higher performance memory and IO interfaces, enterprise system management features, and other features to manage power.

Table 5.1(d) shows the top server microprocessors at roughly five year intervals from 2005–2020 as rated by the PassMark benchmark and website. For the first entry in the table, we took special note of the first multicore processor, in 2006 rather than the top 2005 processor. (All these results are from the database on the PassMark site, https://www.cpubenchmark.net/. The PassMark multicore benchmark is based on throughput, executing independent PassMark single core benchmarks on each hyperthread on each core.) Microprocessor performance improvement in this period shows a 48× speedup. This table can be compared to Table 5.1(c) of TOP500 clusters in terms of relative performance gain every 5 years where clusters gained a 1,000× speedup in this period. Here are some other growth factors for microprocessors between 2006 and 2020: cores 32×, cache 128×, memory bandwidth 38×, and TDP 7×. It is interesting to remark that the memory bandwidth increase has been achieved by the evolution of the memory standard interface from DDR2 to DDR4 as well as ramping-up the number of memory rails. Recently, DDR6 and HBM2 are available.

In Section 5.7, the dark silicon study will be discussed. It models the practical limits of multicore scaling [EBAS11]. A factor not considered in that study is the surprising TDP increase by a factor of 7. But one should add that Table 5.1(d) shows the highest power packages that are used in specialized high cooling environments. One infers that cooling devices have become more prevalent and aggressive.

Another dimension of recent microprocessor evolution is RISC versus CISC. For many microprocessor examples in the book we have used x86 processors. Today, there are two main lines of multicore microprocessor architecture. The first of course is the x86 architecture. X86 is still CISC and it consistently adds (or sometimes drops) instructions. For example, in 1998 AMD added 3DNow! instructions, which were dropped in 2010 as other vector extension instructions superseded them. Perhaps more interesting, x86 is "democratic" with extensions to support all types of computing. For example, x86 has added instructions for security and encryption as well as graphics, AI, and HPC.

The second main multicore architecture is ARM. Although there were quite a few RISC architectures used in supercomputing from the 1980s until 2000, hardly any have survived. ARM remains and it is focused on multicore processors for the datacenter with extensions to HPC, notably Scalable Vector Extensions (SVE). ARM microprocessors are recently very strong in multicore.

The history of ARM is fascinating. ARM started in 1983 in the UK meaning Acorn RISC Machine. It became ARM Ltd. in 1998. Then, following just one of many storylines StrongARM was developed by DEC and sold to Intel in 1997 to replace the i860 architecture. Intel changed marketing strategy to Xscale around 2002. Xscale PXA, an SoC device, was sold to Marvell in 2006 who has come back with ARM as a datacenter part. Another thread is with Apple; the A12 Bionic is ARM architecture.

Here are a few recent ARM multicore processors.

- Marvell Thunder X3: 96 ARM cores each with 4way SMT … that is 384 threads per socket.

- AWS Graviton2: AWS refers to Amazon's AWS clusters in the cloud. It has 64 cores of 64-bit processors.

- Ampere ARM: 80 64-bit cores that are 4-wide superscalar with out-of-order execution, as well as two 128-bit SIMD processors on the chip.

Here are a few recent ARM cluster data points.

- Cray XC50: ThunderX2 blades have 192 nodes per cabinet with dragonfly Aries interconnect.

- Cray-Fujitsu CS500 with 48 A64FX cores, having SVE 512-bit vectors.

- A64FX Fugaku cluster based on 48 core A64FX SoCs is currently Number 1 on the TOP500 list at 415 petaflops.

- Sandia National Lab's Astra cluster with Marvell ThunderX2 processors was formerly the top ARM processor cluster in the TOP500 at 2.3 petaflops. Sandia recently announced acquiring an A64FX cluster.

Although, in Section 5.7, we note that [HePa19] argue that RISC architecture simplifications have advantages over x86, today ARM is approximately the open architecture equivalent of the x86 server line. Which is to say it too faces: the tapering of Moore's Law discussed in Section 5.6; the dark silicon effects due to power limits; and the proposed decline of general-purpose processors in favor of special-purpose processors [ThSp18].

5.5   GPUS—VECTOR ARCHITECTURES UNLIKE PIPELINED SUPERCOMPUTERS

The GPU is the dominant microprocessor architecture in graphic processing for both consumer and professional graphics. They are now also used in datacenters for HPC, with different core configurations. Compared to today's multicore microprocessors, GPUs eliminate a great deal of the control logic in instruction processing and reduce cache size to pack more ALUs onto the die. The GPU architecture, such as Nvidia's, echoes architectural concepts from the history of parallel processing. Even though the implementation of GPUs is very different from previous parallel processors. (It is important to note architectural concept similarities for software developers.) Here are several examples.

- SPMD: The GPU programming model is SPMD with fine-grained threads, *micro-threads*. In this model, each thread is given a unique thread index so that the program can compute a pointer to its unique data items. Compare that to MPI, which is also SPMD notation but implemented at the process level. They use the same work identification model, but greatly different programming languages. The GPU goes further in that the collection of microthreads, called a *grid*, is broken into *blocks*, which are broken into sets of 32 threads, called a *warp*. Warps are the units of microthread work issued in the control unit.

- SIMD control units: GPUs have many control units, called Streaming Multiprocessors (SMs), than MPPs had. Nvidia Ampere GA100 has 128 SMs. SIMD arrays had fewer control units, each controlling a set of PE, e.g., the TMC CM-1 and CM-2 had eight subcubes, each with its own SIMD control unit.

- Branch divergence: In both GPUs and SIMD arrays, all microthreads in a warp or PEs in MPPs executed all instructions including those on both sides of a conditional. (However, Nvidia Kepler introduced dual instruction issue units, so if both sides are independent, they may execute simultaneously.)

- Warps: Although the implementation could not be more different, GPUs have 32 threads per warp that at a conceptual level compares to the vector register on the Cray and AVX registers on the x86 architectures in that the programmer should keep these sizes in mind for most efficient execution.

- Shared random access memory: GPUs have a shared memory random access model like multiprocessors. This was of necessity to allow GPUs to make local changes in image buffers.

- SIMT: Single Instruction Multiple Thread was introduced with GPUs. On multicores, the threading model goes back to Unix System V's "light-weight processes". (Despite the name, this was a threading model.) However, GPUs have *microthreads* that are supported directly in hardware.

- Host processor/Attached processor model: GPU languages use the model in which the program starts on the host that does a memory move to transfer data to the GPU. The host does a remote call on the GPU to run the computational kernel. Finally, the host transfers the results back. This model was used on attached processors like the Floating Point systems AP-120B and on MPPs.

In Section 2.5.5, the evolution of several generations of Nvidia GPUs are compared. There are also different types of GPUs in the same generation now. For example, Nvidia provides: Turing for graphics systems, Volta for datacenters, and Xavier for embedded systems such as autonomous vehicles. Table 5.1(e) compares recent datacenter and consumer versions GPUs. Datacenter versions have almost twice as many cores and consume more power as a result. These usually use HBM2 memory cubes versus consumer versions that use less costly DDR6. New GPUs such as Nvidia Ampere are expected in 2020. Beyond Ampere, Hopper is reported to be in development. Hopper is said to be preparing for MCM implementation to keep the rate of scale-up high as Moore's Law is eclipsed.

Table 5.1: (a) 20-year revenue comparison of Intel, Cray Inc., and Nvidia at 5-year intervals

	Intel (INTC)		Cray (CRAY)		Nvidia (NVDA)	
	Revenue ($B)	Growth (YOY)	Revenue ($B)	Growth (YOY)	Revenue ($B)	Growth (YOY)
2000	29.39	11.86%	0.118	NA	0.37	136.67%
2005	34.21	13.50%	0.149	-8.80%	2.01	10.26%
2010	35.13	-6.54%	0.284	0.10%	3.33	-2.87%
2015	55.87	6.00%	0.562	3.60%	4.68	13.37%
2020	71.96	1.58%	0.456	6.30%	10.92	-6.81%

Table 5.1: (b) Descriptions available for a sample of deep neural net processors

Company : IC Name	Some Key Differences Described	Core Per Chip	Core configuration	Control Unit Features	Memory (MB)	Internal Interconnection	Inter Chip Communication
Cerebras : WSE	Wafer scale for high-speed communication	400,000 SLAC/wafer	SLAC = Sparse Linear Algebra Core	Cores individually program-mable	400,000 GB	Swarm 2D array	100 Pb/s
Mythic : IPU	Low Energy Deployment (0.5 pJ/MAC)	~120 Tiles	Tile = 8x8 8bit Analog MatMul, DSP, SRAM, Router	RISC-V	~120 MB internal	2D array	4 PCIe
Groq : TSP	Totally Systolic VLIW Batch = 1	320 Lanes	Lane = 18 MACs, 2 Perm, 2 Mem, 16 FP16 ALUs	VLIW totally determinis-tic execution	220MB internal	1D bidirectional systolic	16 PCIe
Habana : Gaudi (Acquired by Intel)	Separate Inference and Training ICs	8 TPC2.0s + GEMM Engine	TPC2.0 = C program-mable core with all data types below FP32 + special DNN function units	VLIW SIMD latency hiding	4 external HBMs (32 GB each)	(not specified)	10 100GE RoCE
Graphcore : IPU	In processor memory	1216 Tiles	Tile = FP32 MAC, 256KB SRAM	6-way SMT Bulk Synch-ronous Parallel	304 MB dis-trib-uted inter-nally	All-to-All Ex-change	10 IPU Links @ 64 GB/s
Wave Comput-ing : DPU (Chap.11 Apr 2020)	Integrated pro-cessors and AI chips	16K DPE + 8K DAU	DPE = MAC, RAM DAU = MAC, Perm	SIMD + MIMD in-structions	External 4 HBMs = 2 DDR4	Hierarchical	16 PCIe
SambaNova Systems (Not An-nounced)	Machine learn-ing problem solving system	(NA) Pro-gram-mable Logic like an FPGA?	(NA) Dynamic precision arithmetic, Sparse neural nets?	(NA) Machine learning lan-guage?	(NA)	(NA) 2D mesh?	(NA)

Table 5.1: (c) TOP500 cluster performance and parameters of the top cluster every 5 years since 2005

	Cores	Memory (GB)	Cache/ Core (MB)	Rmax (linpack TFLOPs)	Rpeak (TFLOPs)	Nmax	Node Bandwidth (Gb/s)	Interconnect Fabric
Fugaku – Fujitsu A64FX 2.2 GHz with 512bit SVE, Tofu interconnect D								
6/2020	7,299,072	4,886,048	8	415,530	513,855	20,459,520	unknown	Tofu interconnect D
Tianhe-2A - TH-IVB-FEP Cluster, Intel Xeon E5-2692 12C 2.200GHz, TH Express-2, Intel Xeon Phi 31S1P								
11/2014	4,981,760	2,277,376	2.500	61,444.50	100,679	9,773,000	160	TH Express-2
Jaguar – Cray XT5-HE Opteron 6-core 2.6 GHz								
11/2009	298,592	598,016	2.667	1,941	2,627.61	6,329,856	160	Cray Gemini interconnect
BlueGene/L beta-System - BlueGene/L DD2 beta-System (0.7 GHz PowerPC 440)								
6/2005	65,536		.256 MB	136.8	183.5	1,277,951	16.2	3D Toroid

Table 5.1: (d) Top performing microprocessors at 5-year intervals since 2006

Date	Pass marks	Power (W)	Clock (MHz)	Cores x threads	Cache MB(L3)	Memory interface	Memory bandwidth GB/sec
14nm AMD EPYC 7742							
3/2020	48,062	225	2250	64 × 2	256	8 × DDR4 3200	204.8
22 nm Intel Xeon E5-2699 v3 @ 2.30 GHz							
9/2014	15,584	145	2300	18 × 2	45	4 × DDR4 1600/1866/2133	68
45 nm Intel Xeon W5580 @ 3.20 GHz							
3/2009	3,551	130	3200	4 × 2	8	3 × DDR3 800/1066/1333	32
65 nm Intel Core Duo T2700							
6/2006	1007	31	2300	2	2*	?	5
					(*L2 cache)		

Table 5.1: (e) Recent datacenter and consumer GPUs

Date	Core Configuration	Clock (GHz) and *Boost Clock*	Single Precision Gigaflops	Mem Type	Mem Size (GB)	Bandwidth (GB/sec)	TDP (W)
				Datacenter: Radeon Instinct MI60			
11/2018	4096:256:- 64 CU *	1500 *1800*	14725	HBM2 4kbits	32	1024	300
				Consumer: Radeon TX5700M			
1/2020	2304:144:64 36 CU	1465 *1720*	7926	DDR6 256bit	8	385	180
	(* Unified Shaders : Texture Mapping Units : Render Output Units and Compute Units (CU))						
				Datacenter: Nvidia TITAN V – CEO Edition (Volta)			
6/2018	5120:320:128:640 (80) (6) **	1200 *1455*	14900	HBM2	32	870.4	250
				GeForce RTX 2080			
1/2019	2944:184:64:368:46 (46) (6)	1380 *1590*	9362	DDR6	8	448	180
	(** Shader Processors : Texture Mapping Units (CUDA Cores) : Render Output Units : Tensor Cores (Streaming Multiprocessors) (Graphics Processing Clusters))						

Table 5.1: (f) Comparison of recent CPU to GPU to DNNP. (Adapted from [JYPP17].)

	CPU (AMD EPYC 7742)	GPU (Nvidia Turing)	DNNP (Google TPU)
Specialization	General purpose	Graphics rendering or GPGPU	Deep Neural Nets
Cores	64 (Float64)	5K (Float32)	65K (Int8)
SMT threads	2	64	1
Vectors	16 @ Float32	32@Float32	>256@Int8
Preferred arithmetic	All data types	Float32 and Pixels	MACs Int8, accumulators float32
Control Unit	SMT	60 Streaming CUs	Basic
	Fully pipelined instructions	SIMT	systolic or wavefront
	Branch prediction	SMT	VLIW
	Speculative execution		
	Out of order execution		
	Multitasking with virtual memory		
Cache or On-chip mem	256 MB (L3 cache)	6 MB (L2 cache) + 18 MB(reg file)	200 MB (on-chip mem)
Mem bandwidth	200 GB/s (spec)	800 GB/s (spec)	34 GB/s (Google TPU meas.)
Peak Tflops or Tops	2.5 Tflops(Float32) (spec)	16.3 Tflops(Float32) (spec)	92 Tops(Int8) (measured)
Ops/MemBw per Spec	12.5 Flop/word (spec)	20 Flop/Word (spec)	2700 Ops/Byte (measured)
Ops/MemBw	102 Flop/Word (E5-2699 Haswell 18 core measured)	70 Flop/Word (Tesla K80 measured)	2700 Ops/Byte (measured)

5.6   BEYOND MOORE—ICS, MCMS, AND SOCS

Technology watchers point out that the rate at which smaller lithography fabrication processes come into production is slowing. Thus, they infer the end of Moore's Law may be approaching. As a result, an increasing amount of research and development in several alternatives to CMOS scaling is in progress. This section looks at three fabrication options being researched: ICs, MCMs, and SoCs. This is a broad subject and in a few pages one can mention only a few directions and references that the reader may pursue to understand how this research may change parallel processing.

A case can be made that with the slowing of Moore's Law, advances in general-purpose computers may also be coming to end. This is based not only directly on fabrication technology but also on the economics of fabrication. In [ThSp18], they point out that from before 1980 general-purpose computers were in a virtuous cycle: IC technology advances ➔ New applications of general-purpose computers appear ➔ More investment ➔ More technology advancements ➔, etc. This cycle has led to fabrication lines that cost over $7 billion. On each new line achieving profitable yields takes more time. They propose an economic model to decide when special-purpose processors implemented with less aggressive processes are cost optimal over a general-purpose processor on the most advanced process. The model is based on the case that with parallel processing a special-purpose processor can achieve speedup over the same computation on a general-purpose processor. In this case, a modest speedup with a special-purpose processor, e.g., 3 times faster, if it is fabricated on a process a couple of generations behind the leading edge, would need a production run of less than 200K devices to be cost optimal. They list ten GPU applications that achieve speedups greater than 3 over a multicore microprocessor, and the GPU implementation may be only the start of what could be done in a special-purpose device.

An optimistic point of view is that special-purpose processors may not be needed yet. The ending of Moore's Law may be an eventuality in that process scaling cannot last forever, but similar to when we thought in the 1980s that Amdahl's Law would put an upper limit on parallelism. Loopholes that may appear that can be put into large-scale production in a few years, such as 3D devices, multi-gate devices, SoCs, or MCMs.

3D and multi-gate devices: 7-nm fabrication is in production in 2020 with FinFET devices. A FinFET is a 3D device with a gate formed on three sides of the channel, so that increased energy efficiency and lower gate delay is achieved. A 3-nm FinFET has been demonstrated and may be in production in a few years. Several multi-gate devices are being researched which may also decrease the effective feature size. Although decreasing the feature size may continue at a slower pace, 3D silicon structures may allow "effective feature size" to continue to decrease. For examples, see [Moor18]. It is likely that with any of these devices similar logic rules would apply. So too similar architecture rules, hence parallelism design principles would be similar.

SoCs: In Section 2.5.5, we pointed out that SoCs are crucial for commodity-scale opportunities for parallelism. On an SoC, it can be hard to quantify the "computing power" of design blocks that are close to the edge between digital and analog. Blocks like networking wireless, telephony, camera(s), (ultra)sound transducers, GPS, IoT actuation, the list goes on, allow one to say the effects of integration scaling is still growing. By that we mean that integrating more functions into the SoC makes the system containing the SoC physically smaller. Each SoC block may implement a different analog interface externally but the digital interfaces internally are usually consistent with the historical digital interfaces. Hence, the designer's view is consistent with multi-functional unit parallelism. This process has been successful on Apple iPhone SoCs where by the 11th generation they contain 44 special purpose blocks and the CPU is less than 20% of the chip area [ThSp18].

In general, the current IC technology roadmap is very rich. See the IEEE *International Roadmap for Devices and Systems* annual series of reports that review research issues across the field. Of particular interest may be: *More Moore* [IRDS20a] that discusses 3D and multi-gate developments; *Beyond CMOS* [IRDS20b] covers FPGAs and some other devices that can be used in building parallel systems; and for storage technologies see [IRDS20b].

Multi-chip modules (MCMs): By mounting multiple dies on a common substrate or connecting them with through-silicon vias, whether it be front-side, back-side, or both, interconnect speed, power, and density are much better than mounting chips on circuit boards [KuNa17]. Several examples have been mentioned above. As with SoCs, MCMs are consistent with current parallelism architectures. In fact, with small enough substrate lithography sizes, e.g., the 7 nm seen today, MCMs provide the flexibility over SoCs of choosing the optimum fabrication process for each system block. Putting system blocks on different dies can also reduce the cost of fabricating one larger die without losing significant performance. An example of this approach is the 64-core Ryzen Threadripper described in Section 5.4. For a discussion of a recent MCM process see [Alco20]. At the cluster node level, MCMs are already having an impact as multiple GPUs are being mounted on the same substrate with the added benefit of all-to-all connectivity, Intel Xe Link, for example. [Shal20] points out the research into optical links mountable on MCMs. The energy consumed by data movement on local connections such as those on the node has not decreased appreciably over time and that current research could provide a major bandwidth boost without significant power increase.

Going beyond CMOS devices is possible but likely to take very large investments. For example, spintronics are one of many of the devices being explored that could scale well beyond CMOS. (Spintronics store information in nanomagnets [WNNC17].) Some new devices could change conventional binary logic and memory rules, which may be good and bad for current parallel architecture. Good in the sense that they may lead to new parallel modes of computing, such as quantum computing, to complement digital processing. Bad in the sense that adapting to new parallelism design modes would add another decade to the one to two decades needed to put these devices into production use. An example of this long delay is that the best application method to use may change as a result of using a new device. In this case changes to the entire stack up to application software is needed [VeDC17].

Non-silicon devices aren't new to supercomputing. In the 1980s Gallium Arsenide (GaAs) devices were researched thoroughly because among other properties their higher electron mobility made gates much faster than ECL gates of the time. In 1992, the PEs in the Fijitsu VPP500 combined GaAs, BiCMOS, and ECL circuits. The Convex C-3 was implemented with low density GaAs FPGAs. The Cray-3 was a GaAs supercomputer launched in 1993. Only one was installed and none were sold before Cray Computer Corp. went bankrupt in 1995. Research in GaAs has continued and some special purpose devices have been manufactured over our 1980–2020 period but there has not been a major breakthrough to high quantity. GaAs history underscores how difficult it will be to launch a non-CMOS ecosystem.

[Shal20] concludes that MCMs with application specific processors, as small as DNNPs or as large as Anton, may be the best choice for progress over the next decade while beyond-CMOS devices may take 20 years to bring to production. While new computation model devices like quantum computing are a decade or more beyond that.

5.7   NEAR TERM—SYSTEM AND SOFTWARE SCALING

In addition to devices improvements discussed in the previous section, architecture researchers have made the case that in this decade parallel system and software architecture may keep us on a scale-up path using current IC technology.

Adapted from [JYPP17], Table 5.1(f) reviews the high-level specifications of three systems on a three-step path through the tapering of Moore's Law: an x86 multicore microprocessor (general-purpose architecture), a GPU (special-purpose evolved into limited general-purpose), and a DNNP (special-purpose architecture). The x86 multicore is the Swiss army knife of computing. However, it may be reaching its limit of lithography scaling. Even the multicore dimension may have limits, which are argued just below. The GPU has demonstrated greater power-performance, but only on scalable computations. The DNNP outperforms even the GPU on performance and power, but, also, only on computations that are more scalable and can be tuned further with special-purpose memory and arithmetic. In Chapter 4, we have seen that there are several highly scalable applications that impact many people. Although it remains to be shown whether they can cross the economic threshold described in the previous section, perhaps some of these applications can scale-up on special-purpose processors. Anton in molecular dynamics is a proof point.

Will multicore scaling, the first step of the three-steps shown in Table 5.1(f), be scalable enough? In 2006, multicore scaling was announced as the next step in microprocessor evolution intended to head-off the limit imposed by the end of Dennard Scaling. Only six years after the release of the first multicore, an analysis was published predicting the limits of multicore growth [EBAS11]. They argued that if the end of Dennard Scaling led to multicore, it would eventually limit multicore growth also. That is, average power consumption would exceed package cooling to the extent adding more cores is not beneficial. They proposed that with manycore either some cores are turned off while running less parallel computations or the power/speed per core for parallel computations is scaled-back so that the total power consumed meets the TDP limit. They coined the term "dark silicon" for the turned off cores. (Microprocessor manufacturers with the increasing layers of core power control that were discussed in Section 2.5.5 already support several dark silicon modes of operation.) Dark silicon research modeled not only CPUs cores, but also GPU cores. It was based on 512 processors manufactured up to that date. They concluded that, for their 12 benchmark workload, at 22 nm 21% of the chip must be powered down to keep power consumption below the current TDP. For 8-nm fabrication 44% must be powered down. In short, [HePa19] estimate that we have had to move on from multicore scaling alone sometimes between 2010 and 2015. (Recent heterogeneous core models, in which large high-power cores and small low-power cores are combined in the same MCM, as with the Intel Lakefield processor, were not explored in these reports.)

To go beyond the end of multicore scaling, here are some of the techniques that have been proposed.

- Specialization: The dark silicon model was based partly on general-purpose processors. However, special-purpose processors have proven to be different. First, note that not every application is suited to a special-purpose processor, far from it. [HePa19] point out four ingredients that allow an effective special-purpose processor: (1) abundant parallelism that can use many core; (2) high regularity with good locality for simpler control units, regular interconnections, and less cache; (3) lower, or, at least adapted arithmetic precision, below 64-bit floating point to go to smaller cores, and (4) macro-dataflow/kernel library like parallel notation to make highly tuned architectures easily programmable. In addition, the cost has to be affordable under some economic model. There have been several cases now, which support these points in varying degree: DNNPs 30x, or more, faster than multicore processors; 180x faster ASIC cluster for molecular dynamics, the Shaw Anton; and the Apple Bionic.

- Simplification: [HePa19] argue that CISC architectures are inefficient. They measured speculative execution where 19% of the instructions executed are wasted on an Intel Core i7 on a workload of 12 benchmarks. CISC architectures increase instructions per cycle in favorable computations, but as with speculative execution, they can cause wasted cycles. Plus, the added control logic burns more power. EPIC multi-issue control units are another example of runtime hardware that can be avoided in special cases with VLIW multiple issue.

Architecture researchers also argue that software tools will enable more scalable performance on current technology.

- Software performance engineering: Software for general-purpose systems has been growing under Moore's Law for several decades. Such software has led to a large possibility for inefficiency that can be reclaimed at points where it matters most. [LTEK20] points out that a 4k × 4k matrix multiply can be made over 62,000 times more efficient by: going from implementation in a scripting language on a single processor; to C code on 18 cores parallel with cache blocking; to execution with AVX vector instructions. (This computation becomes 380,000 times faster with a tuned GPU version.)

- Methods and algorithms: [LTEK20] point out ways in which new applications lead to new methods, which leads to new algorithms with large performance gains. In addition, new methods and algorithms can have large economic and end user impact in solving new problems.

- Big components: [LTEK20] also argue that big software components, over 1 million lines of code, such as database engines and websites have offered large opportunities for performance improvement through system tuning.

Finally, one must apply well-proven design principles to achieve good scale-up. To point out one of them consider bandwidth balancing that states that system components that communicate must be sized so that their input/output bandwidths are comparable. If not, performance of the faster components will be limited by the slowest component. For an IC example, [JYPP17] analyze operating point bandwidth using the roofline method for their benchmarks. For a cluster example see [VSBG18]. Defining a cluster is a co-design effort between: the hardware provider, the application developers, and the users. They show that when CPUs were combined with GPUs on a cluster node in mid 2017, e.g., the Titan cluster at Oak Ridge National Lab, one multicore and one GPU were considered a balanced combination. On newer clusters like Summit, operational in late 2019, one finds two or three GPUs per multicore processor. Over time, more applications have utilized GPUs making the new balance point appropriate. ([ThSp18] point out that GPUs increase flops per watt 5 times over CPUs and that the diffusion rate of GPGPU has been about 400 applications over 7 years.) Finally, [VSBG18] note that bandwidth between the node and the interconnection network tends to be further out of balance over time requiring greater locality in the applications and greater memory on the nodes. In addition to bandwidth balancing, all of the researchers cited here point out the importance of co-design. See [JKMM18] for a detailed example of co-design techniques using bandwidth-balancing principles.

CHAPTER 6

Concluding Remarks

The major conclusion of the foregoing discussion is that the growth in parallel processing between 1980 and now has been tremendous, and sometimes not in directions that were clear in 1980. Below are 3 perspectives on the last 40 years: a hardware perspective, a perspective on programming notations, and a perspective on applications.

6.1 A HARDWARE PERSPECTIVE

For an effective parallel computation ecosystem one could say that three components are needed: data, software, and hardware. Here we use broad definitions of these components: data meaning good quality input information in computer-readable form and a lot of it; software including parallel programming notations that are broad and readily diffusible, applications that can be ported to more parallel systems, and run-time environments tuned for the system; and hardware including all components of the system in a form that can enable the software to speedup the processing of data.

During the 1980s, hardware was the driving force. There was a rapid succession of faster computer systems, predominantly faster vector supercomputers. But software and data weren't quite ready. Let us denote that as: hardware → software → data, i.e., "hardware drove software, which in turn drove data." In the 1980s, venture capital to build hardware, i.e., new parallel systems, went beyond software and data available to process on them. When we say software wasn't ready, there was considerable research into new, potential programming notations, much more than today, but there were few production applications to use it. Seismic and FEA applications, yes; genomics, chemistry, were all research, no industrial production; graphics, vision, speech weren't ready to seize multi-million dollar supercomputers. Regarding data, five of the seven applications in Chapter 4 did not show significant growth in data until into the 1990s. Also, note that six of the seven applications appear to be able to grow into the future.

Cray supercomputers were rapidly accepted by large R&D labs in the 1980s. Hitachi, NEC, and Fujitsu were also successful in the Japanese large labs sector. R&D labs had the budget and skilled engineers to buy big machines and develop new applications. Mini-supercomputers, Alliant, Convex, and more, aiming at smaller R&D labs in the industry sector, didn't find a sufficient number of them interested in following in the footsteps of the large labs. Without producing a great number of results to impact many people, after a few years, available venture capital for parallelism decreased. Massively parallel processors stayed in research mode. The parallel hardware in SIMD machines with potentially thousands of processors, notably Thinking Machines and MasPar, led the way to MPPs. They were built because of the potential application performance but complete applications trailed too far behind to sustain these companies.

After the hardware-led boom in parallel processing, there was a period of recovery. The 1990s and 2000s was a period when software growth led the way. (software → hardware → data.) Linux, being open source, took the custom OS out of the hands of hardware system companies and made large distributed clusters more affordable to many. RISC killer micros were well suited here because they were already running Linux on workstations!

The programming notations software community was well suited to seize the lead too. DMM message passing and shared memory primitives had each diffused to healthy size communities that had produced different implementations. With the open source movement, they realized the opportunity to standardize to enable application developers to move forward. The open source model also fueled rapid diffusion of new parallel applications to the point that at least five of the seven application fields discussed in Chapter 4 had, and still have, major open source packages.

Parallel hardware followed software's leads. Small-scale SMM systems built with RISC microprocessors were a straightforward extension of workstations and servers for SGI, Sun, Digital, HP, and IBM. They made the easiest architecture to parallelize applications for. Parallel applications for DMM systems required much more time to program. So, the companies still around after the supercomputer boom stepped in with NUMA RISC systems, notably: SGI Origin/Altix and the HP-Convex Exemplar; while Digital and Sun chose to back others: Cray T3D, TMC CM-5 respectively. DMM clusters built momentum as message passing broadened and diffused. The MPP hardware vendors supported cluster growth by producing the interconnection networks: (more or less successfully) Intel iPSC, nCUBE, TMC CM-5.

In the software → hardware → data period, parallel programming notations and applications matured sufficiently to enable hybrid parallel architectures starting with DMM systems containing SMM nodes such as ASCI White. At first there was no change in the programming model and application developers just ran multiple MPI processes on the SMM nodes, but hybrid programming paradigms became increasingly popular in time. (SMM-vector hybrids were already successful in the hardware-led period, e.g., Cray X-MP, Alliant FX-8, Convex C-2.)

Around 2000, data started taking the lead from software; we were entering the data → hardware → software phase. Several applications accumulated data continuously so there was no precise date when this occurred for them. Seismic processing was one application that had steadily increased the size of geophysical survey data over 40 years and seismic data processing centers were a consumer of DMM clusters such as the Cray XC30 [Bran17]. Some applications just needed time for human experts to build larger, more accurate models, e.g., FEA, whole car models assembled from each car part, and CGI, animation character models detailed enough for realistic body and facial feature motion guided by kinematics. Data storage prices continued to come down continuously driving the need for clusters in these applications.

But, around 2000 the genomics and computational chemistry labs started producing large amounts of data. As was noted in Sections 4.4 and 4.5, DMM clusters work well for these applications.

By 2010, data collection on the Internet was increasingly driving applications like speech recognition. In addition, less expensive data collection devices, e.g., solid-state cameras, running in video mode began to drive computer vision. Predominantly Internet data driven companies like Google, Baidu, Amazon, and Microsoft began to be interested in clusters as supercomputers with the emergence of machine learning and deep neural network applications. Parallel hardware followed with GPUs.

GPU manufacturers reacted with Tensor Processing cores. FPGAs for use in datacenter clusters also appeared, e.g., Microsoft Brainwave. Data leading hardware also worked for personal devices. This effect can be seen in the design of Apple's Bionic and the addition of AVX512-VNNI to the x86 platforms (vector neural network instructions).

At the high-performance computing level, the applications used on the TOP500 clusters started including data centric applications so that memory and IO started to grow in clusters. For example in 2010, the Graph500, a non-numeric benchmark composed of: building a large graph with sparse data structure, parallel breath-first traversal, and single-source shortest path computation, appears (although there aren't 500 system submissions at that time). By 2013, there were nine IBM Blue Gene/Q systems in the top 11 spots in the Graph500. Increasingly, this became an international phenomenon; in the 2019 at the top of list one finds: TaihuLight, Tianhe-2 clusters from China; IBM and Lenovo clusters in Germany, Italy, and France; a Cray cluster in Korea; as well as the perennial U.S. labs, LLNL, Argonne, ORNL with IBM clusters, and NERSC with a Cray system. Finally, starting around 2010 DNNP companies started up, as we saw in Section 5.2.

It is hard to say how long the data driven phase will last. A philosophical point of view is that this phase is economically stable because data is an eternal, customer oriented driver whereas hardware and software are technology drivers. If one looks at the previous phases, they were each dominant for 15 years and data has been dominant for 10 years so far.

6.2   A PROGRAMMING NOTATION PERSPECTIVE

In terms of programming notations, we can say that the last 40 years gave us the powerful notations of today for distributed memory machines, shared memory machines, vector operations, and GPUs. The germs of all of today's programming notations were ideas already available in 1980. For example, as mentioned in Chapter 3, message passing notation was introduced in the literature with the CSP notation, in the late 1970s, years before message passing was used to program DMM systems. Shared memory programming notations are rooted in tasking primitives and parallel loop constructs introduced before 1980. And GPU programming notations can trace their origins to ILLIAC IV's Glypnir. However, programming notations, especially those for distributed and shared memory programming, have gone from a minimal orthogonal set of constructs to an extensive library of functions, and from short descriptions to today's specifications and books containing hundreds of pages to help new users acquire parallel programming skills.

A second observation is that programmers of parallel machines are extraordinarily demanding to the point that 10% performance loss has often been considered unacceptable. Today's compiler technology cannot guarantee generating code from high-level programming notations, such as array operations, that is within 10% of what can be obtained by a programmer with lower level programming notations. Although there is little experimental evidence in the literature, it is common lore that the code generated by today's compilers, even sequential language compilers, is typically suboptimal and sometimes significantly suboptimal. Most programmers are not aware of this since it is not easy to notice that there is much room for improvements in the case of sequential systems, but for parallel machines, speedup measurements give away compiler inefficiencies. Thus, while there is no natural upper bound to the improvement that can be expected from the application of sequential optimizations, the number of processors and the number of pipeline stages are a natural upper bound for speedup. When the compiler delivers a speedup that is significantly less than the number of processors it is a trivial and quite clear indication that the compiler failed to deliver an efficient executable. The end result of today's compiler limitations is that MPI, categorized by some as "the assembly language of parallel programming," is the norm for programming high-end machines instead of a higher-level programming notation. A similar situation arises in the case of the SIMD vector parallelism of microprocessor vector extensions and in the case of GPUs where lower-level intrinsics notation and SPMD programming are used instead of array notation. Also, for many application domains, library notations are preferred to programming languages. Although a library notation is more verbose and compile-time error detection is not as powerful, the use of libraries does not support compiler analysis, which would greatly facilitate the implementation, portability, and evolution of programming notations.

Looking toward the future, some less efficient but safer scripting programming notations may gain more traction. Parallel hardware has become orders of magnitude less expensive and more available so that programmers more often change their attitude from "the programming notation must provide efficiency" to "if some part of the application is slow, how can I use the available hardware without expending a lot of programming effort?"

The parallel programming community has been quite successful at coming together to create standard programming notations that do not change the underlying programming language, e.g., APIs, templates, and directives. Although none of the major programming notations has been standardized by ANSI, as has been the case for most of the popular conventional languages, the community has been able to form organizations to regulate programming notations that have received the support of academic and industrial organizations. These organizations include the OpenMP Architecture Review Board, the MPI Forum, and the Khronos Group.

The landscape of parallel programming notations has been stable in the recent past with MPI, threading, OpenMP, TBB, CUDA, OpenCL, etc. However, these notations continue to evolve in the form of new extensions and implementation releases. Thus, tasking mechanisms, SIMD directives, and accelerator directives have been added to OpenMP in the last decade. The MPI forum is considering extensions to better support hybrid programming models and to support fault tolerance among other new features.

Looking toward the future, an appealing programming notation is macrodataflow. As mentioned above, it has been explored as a programming model for future exascale machines and has been incorporated in a few popular systems such as TensorFlow and as an extension to Charm++ [HuKa07]. Although the dataflow notation can complicate program readability, its popularity seems to be slowly growing. There is also renewed interest in array notation for parallelism, in part due to its use in programming notations such as TensorFlow. In part, these trends are due to the massive amounts of data now being processed, which gain significant caching and I/O efficiency by organizing and staging the data for efficient processing. There is also great interest today in domain specific programming notations as a mechanism to facilitate parallel programming. Finally, there is also great interest in programming notations that guarantee determinacy of shared memory parallel programs [BADA09] due to the debugging difficulties created by non-determinacy.

6.3   AN APPLICATION PERSPECTIVE

The seven applications in the book are just a sample showing that parallel applications have achieved breakthroughs in several fields. We showed that through parallel applications many people's lives are better in several ways either directly or as a result of engineers and scientists using the results of parallel computation. To summarize parallel applications, let's note where applications stand with respect to the five themes of the book.

Performance improvement: Many parallel applications have speeded-up as parallel hardware has scaled-up. "Application" here has two senses, i.e., application programs make up an application field. To make the application field successful effort has been put into parallelizing not only the most computationally intense application programs but the application programs with great benefit to the application field. For example, drug design applications are pleasantly parallel and easy to put on scaled-up hardware. Molecular dynamics is also parallel but it gets harder to improve as the hardware scales-up.

Popularization: Recently, applications run on commodity devices have been parallelized to directly impact end users. Computer vision and speech recognition are two recent examples, starting around 2010. Computer-generated images scaled-out earlier. In 1995, Sony invented the acronym

GPU for their polygon-rendering engine when it was put into the Playstation. The same year, with ray tracing, the entirely computer-generated Toy Story came out.

Persistent incremental innovation: Innovation in parallel applications has been a consistent story of parallel hardware inspiring new applications and parallel applications inspiring new hardware. For example, DYNA3D was inspired by the vector processors of 1980. By 1990, the opportunity to make computer games more realistic inspired GPU hardware. By 2000, a cluster of SMM systems allowed the human genome to be sequenced, which has inspired considerable parallel hardware at all levels. The field of GPGPU applications grew from GPU performance increases. In 2005, LU factorization became reality on a GPU. Which in turn changed the composition of clusters allowing them to reach a higher level of parallelism. As mentioned in the hardware perspective, the 2010 decade has seen remarkable innovation between DNN applications and DNN processors. Today and looking forward, new application fields are inspiring new application programs in bioinformatics and computer vision, to name a few, and new hardware such as exascale clusters and beyond-Moore technologies are coming out to sustain application innovation.

Broadening and diffusion of programming notations: As noted in concluding remarks for programming notations the focus is on supporting the main architectures and enabling the application programmer to get directly to the parallel performance those architectures offer. This and a mistrust of optimizers is not surprising to application developers. When one works on the same application for 10–20 years or more, one knows where and how parallelism should be used. Supporting this simple, direct approach, the clearer the programming notation is, the easier software maintenance is.

In each parallel application subsection, we showed that the applications became available to run on the major architectures when they came into production and even on the hybrids composed of these architectures. That happens because programming notation standard groups and forums have been attentive to application demands and have broadened the programming notations in support as large a range of applications as possible.

Dramatic increase in the number and performance of applications: Throughout the book we have pointed to the drivers of application scaling: data size growth, new parallel methods, new scalable layers, and time-to-solution pressure. Data size growth is driving just about every application category discussed, 6 of 7. New parallel methods and additional scalable layers occur just about as often, 5 of 7 each. Although time-to-solution pressure doesn't occur as often (only 3 of 7 times), it is likely it will appear more as more applications use the growing parallelism in commodity devices. These scalability drivers have led applications to follow several parallelism scaling patterns over the 40 years: high growth over the whole period, growth leveling-off as the fundamental problem is solved, growth shadowing or track client device parallelism in the case of direct end user impact, growth migrating down to client devices to enable the application to reach more end users, and new rapid growth fields driven by breakthroughs in solving real-world problems. It is important to note that only one pattern, performance leveling-off, indicates that application parallelism will not continue to grow as quickly as parallelism in hardware grows.

6.4   CONCLUSION

Two questions were frequently asked in creating this history. First, what caused events to occur, and second, what lessons can be learned from them. Upon finishing this document we feel a bit closer to having answered what led us to the dominant parallel architectures and applications we see today. For the second question, when future technologies arise they will most likely invalidate some of the things learned in parallel processing history. However, we expect some of the things learned over forty years of parallel processing will benefit computer science when new technologies emerge in the future. We hope that the reader will abstract from details reviewed here to see analogies to situations they find themselves engaged in. We ourselves hope we can work with others to find and form useful meta or macro observations.

Appendix A: Myths and Misconceptions about Parallelism

To make parallel processing an everyday technology several conceptual challenges had to be overcome through the recent decades.

- "Supercomputer Mystique": Parallel processing was hampered by the availability of parallel hardware for regular users until multicores and GPUs became readily available in the decade 2000–2010. Today, due to broad availability of parallel hardware, every application developer can take advantage of it.

- "People Don't Think In Parallel" and "Synchronization Is Tricky": Parallel programming has been hampered by non-parallel programming notations and application architects that were slow to incorporate parallelism. Indeed in developing a new application, one may write and debug in serial before adding the complexity of parallelism. However, several language designers and application architects were afraid to give their programmers access to parallel constructs assuming that their use would create hard to trace or reproduce problems. These attitudes still exist today. Some software architects say that parallelism is to be used only after all serial algorithmic performance improvements have been exhausted. Now, there is growing familiarity with parallel programming, but solid understanding is still not widespread. On the positive side, with the broadening and diffusion of parallel notations the obstacles keep slowly decreasing. For example, parallel library notations (see Section 3.7) are prevalent in some application domains. Another positive concept is implementing the pyramid of parallel software development team. At the top application architects encapsulate and manage parallelism allowing more programmers, at the base of the pyramid, to benefit from parallelism without having to write parallel code.

- "Amdahl's Law" will asymptotically limit speedup: Despite earlier positive analyses of available parallelism [KuMC72], Amdahl's Law was often cited as why parallel processing was bound to fail. Loopholes to this law [Wolf15] have made the opposite outcome possible. In fact one could say that the main task of parallel processing is to break Amdahl's Law in every practical case. A principal loophole that has occurred in the history of parallelism is data growth over time. That is, the amount of data that applications process tends to increase in concert with the hardware capacity to process the data. Data growth enabled speed-up, sometimes called "scaled speed-up", is often called "weak scaling" where Amdahl's Law describes "strong scaling", meaning speeding-up an application when the data size is fixed [Gust88]. Several of these cases are described in the applications chapter: finite element analysis, seismic processing, image processing, computational chemistry, and bioinformatics. For these cases and many others data growth has been several orders of magnitude since the 1980s.

- "It's For Multitasking, Not Multiprocessing": When multicore microprocessors arrived many thought they would be used for multitasking several applications. But, most users find it difficult to conscientiously keep foreground and background tasks active simultaneously, let alone 4, 8, or more tasks. With respect to multitasking, the number of processes running on an average laptop today is several hundred and a reasonable number can run autonomously, as soon as triggered by the arrival the right data or signal. The number of autonomous services may be increasing slowly over time. With respect to multiprocessing, we are reminded that shortly after the launch of multicore microprocessors, Intel and Microsoft launched a research project at a few of the top universities doing parallel processing research [TAAD13, ABCG06]. Few new killer applications are clear yet and results from this research have been slow in reaching production applications. But gradually or in rapidly-blooming areas, e.g., deep neural nets, one cannot deny that the parallelism is exploited in numerous applications that now impact many people.

- "Most Users Spend Most Of Their Time In Applications That Aren't Parallel": Many widely used features of common applications such as: mail, word processing, spreadsheets, and social media are inherently serial. A serial text stream is a very meagre amount of data to process these days and that limits the potential of today's multicore microprocessors. Gradually more compute intensive features, some of which are mentioned next, are added to these applications allowing more parallelism to take root.

Despite the skepticisms about parallelism above, one should consider a few success stories.

- For processing serial text streams, if one adds features such as speech-to-text dictation, the computational intensity and value of parallelism increases.

- Many thought that games were inherently serial given they must follow the sequential actions of the player. That part is true, but many games feature multiple agents that act in parallel through the game's AI engine for example. The physics engine also frequently can run many different bodies moving in parallel.

- Video data processing, although its use increases slowly, offers much parallelism potential. A now common example is HD streaming video decoders which were futuristic in 1981. By the 1990s, mini-supercomputers were being sold to video research labs. Now, thanks to higher clock rates, more parallel functional units, and microprocessor vector instructions, video decoders use only a portion of one core of today's multicore processors.

- A final example to show that parallelism and perseverance are almost synonymous, many think that browsers are not parallel, and indeed, a great deal of the wait time is network delay. However, browsers do get faster. On the hardware side, browsers implicitly will use different processors on different pages and can use GPU parallelism, e.g., Chrome. Parallelism in browsers use can be expanded further. As page content grows, so does parallelism opportunity. Thorough studies of parallelizing rendering in the browser showed promise [JLMA09] and action is being taken to bring this to production [Clar17]. In the future, as web content becomes more active and complex, studies [RHSD17] show threading capability, such as Web Workers supported in HTML5, may bear fruit.

- With respect to notations, many IDEs now contain features supporting the development of parallel applications, especially the Microsoft and Intel IDEs.

Appendix B: Bibliographic Notes

Here are some bibliographic observations on parallel processing in this period:

- This book focuses on the broad interaction between hardware, notations, and applications that has led to the success of parallel processing in both datacenter servers and client devices. There are more detailed technical analyses of this three-way interaction. See for example [JKMM18].

- The Gordon Bell Prize [BBDK17] is an annual award for outstanding advances in high performance computing. The prize was founded in 1987 and there are now many interesting advances to read about.

- In the late 1980s, the U.S. Office of Science and Technology Policy created a list of research challenges for high performance computing. Their findings were published in 1992 and became known as *The Blue Book* [FCCS92]. The National Science Foundation updated this list and published their findings in 2011 [ACCI11]. Before and after this other countries created HPC roadmaps also, for example, the Japanese Fifth-Generation Computer Systems Project started in 1982. See also the High Performance Computing Advisory Council's website [HPCA18], started in 2009.

- There are several interesting articles on the history of high-performance computing and parallel computers. For example, [LLNL02] covers the first 50 years of computing at Lawrence Livermore National Lab. And, many improvements have emerged since then.

- There are also several interesting informal webpages worth reviewing.

  ○ The TOP500 is a site that lists news about the top 500 supercomputers in the world as ranked by performance on a HPLinpack benchmark. It also contains a database on performance by date that can be analysed from many points of view [TOP500].

  ○ Wikipedia has pages about each of the "World's Fastest Computers" with a link to what was the prior fastest system, when this system became number one on the list, and a link to the system that replaced it. The earliest computer on the list is the Whirlwind I of 1951 [WikWHI]. The most recent on the list, at the time of writing, is the Summit system [WikSUM].

- ○ The WaybackMachine, an Internet archive contains many webpages and documents on the history of parallel computing. For example [https://web.archive.org/web/20041120084657/http://arrakis.ncsa.uiuc.edu/ps2/] is a page about a cluster built in 2003 with Sony PlayStation2s. This little known system was precursor of the GPGPU movement.

- ○ "The History of the Development of Parallel Computing" webpage gives a concise parallel processing timeline. Unfortunately it stops at 1994 [Wils94].

- ○ The "Overview of Recent Supercomputers" webpage is hierarchically organized by architecture type [StDo04] and contains some nice historical insights.

- There are numerous journals and conferences that treat research in parallel processing. Examples include the *ACM Symposium on Principles and Practice of Parallel Programming* (PPoPP), the *IEEE International Parallel and Distributed Processing Symposiums* (IPDPS), and the *European Conference on Parallel Programming* (Euro-Par).

- Those interested in the history of computers should visit one of the many computer history museums. Wikipedia has a page listing computer museums around the world and often the museum's webpage will note if they have parallel computers [WikLCM].

- The *Encyclopedia of Parallel Computing* [Padu11] is a good source of information about a broad spectrum of parallel computing topics.

- In several cases, references in this book cite web pages. We have found this important especially where technical papers have not yet been published, for example in noting trends and directions in parallel applications. We felt that to put past achievements in parallel processing into perspective we should cite some material to discover what direction parallelism may take in the near future.

Appendix C: Taxonomic Notes

Machine organization concepts have remained fairly consistent over the years. The term SIMD (Single Instruction Multiple Data), introduced by Flynn in 1972 [Flyn72] to describe, "processors characterized by a master instruction applied over a vector of related operands." SIMD operations can be executed in pipelined form or simultaneously on multiple identical devices. When discussing hardware, we will use the term vector with the pipelined form and SIMD to refer to processors containing an array of processing elements. Today, in a system where multiple independent processors can dynamically couple together to execute a single instruction and then uncouple, sometimes called Single Instruction Multiple Thread (SIMT) fashion, Flynn's taxonomy becomes blurry. With the large scale of integration available now, replicating control units that operate in this way is practical and effective. Simultaneous Multi-Threading (SMT) is different. In SMT, a control unit keeps register sets for several threads ready so that if an executing thread stalls the control unit can switch to another thread, most frequently to hide memory latency.

A second term introduced by Flynn was MIMD (Multiple Instruction Multiple Data), which we do not use in this book, was a classification for either SMM machines or DMM machines. Machine organizations involving hybrid SIMD and MIMD designs have become pervasive by virtue of Moore's Law applied to microprocessor real estate (see Subsection 2.5.3 and Figure 2.9). The two widely used examples of these hybrid designs are today's multicores and GPUs nodes whose emergence contributed significantly to the popularization of parallelism. GPU technology was first a special purpose processor for graphics, but soon evolved into useful general-purpose computational devices [DWLT12].

Appendix D: The 1981 Tutorial

The 1981 tutorial was a collection of papers that were selected to present a comprehensive view of what was then a nascent field. The first chapter, an introduction, contained two classifications of parallel machine organization and a survey of multiprocessor designs. The next four chapters contained papers describing machines, architecture issues, and parallel applications for each of the four classes of machines covered by the tutorial. Chapter 2 about *pipelined vector processors* contained papers on four machines (CRAY-1, STAR-100, CDC 205, and Floating Point Systems AP-120B) and two applications (Finite Difference Equations and Plasma Simulation). Chapter 3 on Single Instruction Multiple Data, *SIMD machines*, covered three machines (Burroughs Scientific Processor, STARAN, and Goodyear MPP), three architecture issues (Collective Operations, Access Alignment, and Interconnection networks), and two applications (Image Processing and Databases). Chapter 4 on *multiprocessors* included papers on three machines (S-1, Cm*, and the Burroughs FMP design), a paper on memory bandwidth, and two applications (speech understanding and atmospheric weather prediction). Chapter 5 on *dataflow machines* included two machines (a dataflow machine design and Denelcor's HEP) and one application (weather forecasting). In addition to these four classes, Chapter 6 was about *special purpose processors* (the Structured Multiprocessor System (SMS), the Finite Element Machine, cellular logic machines for image processing, and relational database machines). Chapter 7 covered *parallel programming languages* (Concurrent Pascal, Communicating Sequential Processes (CSP), an array language, and Dataflow Languages). Chapter 8 contained three papers on *automatic parallelism* (using automatic analysis to measure the parallelism of ordinary programs, a compiler/runtime system for the exploitation of low-level parallelism in Cm*, and Linpack timings using different compilers including the Cray-1 vectorizing compiler). Chapter 9 covered *operating systems* (for Cray-1, STARAN, and Cm*). Finally Chapter 10 contained papers on *parallel algorithms* (a survey of numerical algorithms, an introduction to asynchronous algorithms, sorting on pipelined processors, and concurrent operations on B-Trees).

References

[AAAB15]   D. Amodei, S. Ananthanarayanan, R. Anubhai, J. Bai, E. Battenberg, C. Case, Jared Casper, Bryan Catanzaro, Qiang Cheng, Guoliang Chen, Jie Chen, Jingdong Chen, Zhijie Chen, Mike Chrzanowski, Adam Coates, Greg Diamos, Ke Ding, Niandong Du, Erich Elsen, Jesse Engel, Weiwei Fang, Linxi Fan, Christopher Fougner, Liang Gao, Caixia Gong, Awni Hannun, Tony Han, Lappi Vaino Johannes, Bing Jiang, Cai Ju, Billy Jun, Patrick LeGresley, Libby Lin, Junjie Liu, Yang Liu, Weigao Li, Xiangang Li, Dongpeng Ma, Sharan Narang, Andrew Ng, Sherjil Ozair, Yiping Peng, Ryan Prenger, Sheng Qian, Zongfeng Quan, Jonathan Raiman, Vinay Rao, Sanjeev Satheesh, David Seetapun, Shubho Sengupta, Kavya Srinet, Anuroop Sriram, Haiyuan Tang, Liliang Tang, Chong Wang, Jidong Wang, Kaifu Wang, Yi Wang, Zhijian Wang, Zhiqian Wang, Shuang Wu, Likai Wei, Bo Xiao, Wen Xie, Yan Xie, Dani Yogatama, Bin Yuan, Jun Zhan, and Zhenyao Zhu. 2016. Deep speech 2: end-to-end speech recognition in English and Mandarin. In *Proceedings of the 33rd International Conference on International Conference on Machine Learning (ICML'16)*, Maria Florina Balcan and Kilian Q. Weinberger (Eds.), 48. JMLR. pp. 173–182. 92

[ABCC16]   Martín Abadi, Paul Barham, Jianmin Chen, Zhifeng Chen, Andy Davis, Jeffrey Dean, Matthieu Devin, Sanjay Ghemawat, Geoffrey Irving, Michael Isard, Manjunath Kudlur, Josh Levenberg, Rajat Monga, Sherry Moore, Derek G. Murray, Benoit Steiner, Paul Tucker, Vijay Vasudevan, Pete Warden, Martin Wicke, Yuan Yu, and Xiaoqiang Zheng. 2016. TensorFlow: a system for large-scale machine learning. *Proceedings of the 12th USENIX Conference on Operating Systems Design and Implementation (OSDI'16)*. USENIX Association, Berkeley, CA, 265–283. 58

[ABCG06]   Krste Asanovic, Ras Bodik, Bryan Christopher Catanzaro, Joseph James Gebis, Parry Husbands, Kurt Keutzer, David A. Patterson, William Lester Plishker, John Shalf, Samuel Webb Williams, and Katherine A. Yelick. 2006. The landscape of parallel computing research: A view from Berkeley. Electrical Engineering and Computer Sciences University of California at Berkeley. *Technical Report No. UCB/EECS-2006-183*. Retrieved Dec. 14, 2019 from http://www.eecs.berkeley.edu/Pubs/TechRpts/2006/EECS-2006-183.html. 76, 130

[ACCI11] National Science Foundation Advisory Committee for Cyberinfrastructure Task Force on Grand Challenges. *Final Report. 2011.* Retrieved Dec. 14, 2019 from https://www.nsf.gov/cise/oac/taskforces/TaskForceReport_GrandChallenge.pdf. 77, 133

[Adam15] Jill Adams. 2015. Genetics: Big hopes for big data. *Nature 527*, S108–S109. DOI: 10.1038/527S108a. 85

[AGMM90] Stephen Altschul, Warren Gish, Webb Miller, Eugene Myers, and David J. Lipman. 1990. Basic local alignment search tool. *Journal of Molecular Biology*, 215(3): 403–410. DOI: 10.1016/S0022-2836(05)80360-2. 84

[AHHH92] F. Alleva, H. Hon, X. Huang, M. Hwang, R. Rosenfeld, and R. Weide. 1992. Applying SPHINX-II to the DARPA *Wall Street Journal* CSR task. In *Proceedings of the Workshop on Speech and Natural Language (HLT '91)*. Association for Computational Linguistics, Stroudsburg, PA, 393–398. DOI: 10.3115/1075527.1075622. 91

[Alco20] Paul Alcorn. August 20, 2019. Intel Lakefield 3D Foveros Hybrid Processors: Hot Chips 31 Live Coverage. https://www.tomshardware.com/news/intel-lakefield-foveros-3d-chip-stack-hybrid-processor,40205.html. 117

[AnCP95] Thomas E. Anderson, David E. Culler, David A. Patterson, and the NOW team. 1995. A case for NOW (Networks of Workstations). *IEEE Micro* 15(1) 54–64. DOI: .1109/40.342018. 51

[Andr20] Mark Anderson. January, 2020. Will China attain exascale supercomputing in 2020? In *IEEE Spectrum*, 57(1): 52–53. 105

[AnTD16] Hartwig Anzt, Stanimire Tomov, and Jack Dongarra. 2017. On the performance and energy efficiency of sparse linear algebra on GPUs. *International Journal High Performance Computing Applications*, 31(5): 375–390. DOI: 10.1177/1094342016672081. 80

[Appe68] Arthur Appel. 1968. Some techniques for shading machine renderings of solids. In *Proceedings of the April 30–May 2, 1968, Spring Joint Computer Conference (AFIPS '68 (Spring))*. ACM, New York, 37–45. DOI: 10.1145/1468075.1468082. 83

[ArKi89] James Arvo and David Kirk. 1989. A survey of ray tracing acceleration techniques. In Andrew S. Glassner (Ed.) *An Introduction to Ray Tracing*, Academic Press Ltd., London, UK, 201–262. 83

[AsLR18] Alán Aspuru-Guzik, Roland Lindh, and Markus Reiher. 2018. The matter simulation (r)evolution. *ACS Central Science*, 4(2): 144–152. DOI: 10.26434/chemrxiv.5616115.v1. 88

[AtSi93] P. M. Athanas and H. F. Silverman. 1993. Processor reconfiguration through instruction-set metamorphosis. *Computer*, 28(3): 11–18. 59

[BADA09]   Robert L. Bocchino, Jr., Vikram S. Adve, Danny Dig, Sarita V. Adve, Stephen Heumann, Rakesh Komuravelli, Jeffrey Overbey, Patrick Simmons, Hyojin Sung, and Mohsen Vakilian. 2009. A type and effect system for deterministic parallel Java. In *Proceedings of the 24th ACM SIGPLAN Conference on Object Oriented Programming Systems Languages and Applications (OOPSLA '09)*. ACM, New York, 97–116. DOI: 10.1145/1640089.1640097. 125

[BaGP62]   F. R. Baldwin, W. B. Gibson, and C. B. Poland. 1964. A multiprocessing approach to a large computer system. *IBM Systems Journal*. 1(2): 64–76. DOI: 10.1147/sj.11.0064.

[BaKi93]   V. Bala and S. Kipnis. 1993. Process Groups: a mechanism for the coordination of and communication among processes in the Venus collective communication library. 1993. *Proceedings Seventh International Parallel Processing Symposium*, Newport, CA, 614–620. DOI: 10.1109/IPPS.1993.262809. 62

[Bane93]   Utpal K. Banerjee. 1993. *Loop Transformations for Restructuring Compilers: The Foundations*. Kluwer Academic Publishers, Norwell, MA. 59

[Batc80]   K. E. Batcher. 1980. Design of a massively parallel processor. *IEEE Transactions on Computers*, 29(9): 836–840. DOI:10.1109/TC.1980.1675684. 16, 28

[BBBC94]   Vasanth Bala, Jehoshua Bruck, Raymond Bryant, Robert Cypher, Peter de Jong, Pablo Elustondon, D. Frye, Alex Ho, Ching-Tien Ho, Gail Irwin, Shlomo Kipnis, Richard Lawrence, and Marc Snir. 1994. The IBM external user interface for scalable parallel systems. *Parallel Computing*, 20(4): 445–462, ISSN 0167-8191. DOI: 10.1016/0167-8191(94)90022-1. 62

[BBBC07]   Rob Baxter, Stephen Booth, Mark Bull, G. Cawood, James Perry, Mark Parsons, Alan Simpson, Arthur Trew, A. McCormick, G. Smart, R. Smart, A. Cantle, R. Chamberlain, and G. Genest. 2007. Maxwell—a 64 FPGA supercomputer. *Second NASA/ESA Conference on Adaptive Hardware and Systems (AHS 2007)*, Edinburgh, 287–294. DOI: 10.1109/AHS.2007.71. 50

[BBDK17]   G. Bell, D. H. Bailey, J. Dongarra, A. H. Karp, and K. Walsh. 2017 A look back on 30 years of the Gordon Bell Prize. *The International Journal of High Performance Computing Applications*, 31(6): 469–484. DOI: 10.1177/1094342017738610. 133

[BeHo18]   Tal Ben-Nun and Torsten Hoefler. 2018. Demystifying parallel and distributed deep learning: An in-depth concurrency analysis. *arXiv*:1802.09941v2. 76

[Bell02]   G. Bell. 2002. A brief history of supercomputing: 'The crays', clusters and Beowulfs, centers. What next? Retrieved Dec 14, 2019 from http://research.microsoft.com/en-us/um/people/gbell/supers/supercomputing-a_brief_history_1965_2002.htm. 40

[Bens07]   David J. Benson. 2007. The history of LS-DYNA. 2007. Retrieved Dec 14, 2019 from https://www.d3view.com/wp-content/uploads/2007/06/benson.pdf. 80, 81, 97

[Bern66]   A. J. Bernstein. 1966. Analysis of programs for parallel processing. *IEEE Transactions on Electronic Computers*. EC-15(5): 757–762. DOI: 10.1109/PGEC.1966.264565. 59

[Boma15]   Karen Boman. 2015. Big data growth continues in seismic surveys. *Rigzone*, Sept. 2, 2015. 79

[Boot18]   M. Booth. Private communication 2018. 65

[Bran17]   Sverre Bransberg-Dahl. 2017. High-performance computing for seismic imaging; from shoestrings to the cloud. *SEG International Exposition and 87th Annual Meeting*. Retrieved Dec. 14, 2019 from https://pdfs.semanticscholar.org/9241/baf4a72ef8b-1821bab97d4c943d225c69336.pdf. 79, 122

[Broo90]   Eugene Brooks. 1990. Attack of the killer micros. Talk at *Supercomputing 1990*. 37

[Burr71]   Burroughs Corporation. 1971. *Array Processing System Fortran IV Reference Manual, Change No.3*. 1971. Defense, Space, and Special Systems Group, Burroughs Corporation, Paoli. PA. 56

[CGNG15]   Shannon Chen, Zhenhuan Gao, Klara Nahrstedt, and Indranil Gupta. 2015. 3DTI Amphitheater: Toward 3DTI Broadcasting. *ACM Trans. Multimedia Computing Communications and Applications,* 11(2s): Article 47, 22 pages. DOI: 10.1145/2700297. 83

[ChFP08]   Srinivas Chellappa, Franz Franchetti, and Markus Püschel. 2008. How to write fast numerical code: A small introduction. Generative and Transformational Techniques in Software Engineering (GTTSE), *Lecture Notes in Computer Science*, Springer, Berlin Heidelberg, 5235, 196–259. DOI: 10.1007/978-3-540-88643-3_5. 18

[ChGK11]   J. Chong, E. Gonina, and K. Keutzer. 2011. Efficient automatic speech recognition on the GPU. 2011. Retrieved Dec. 14, 2019 from http://citeseerx.ist.psu.edu/viewdoc/citations;jsessionid=DE08B46A48E292C5FF2DA89C4E26FFE0?doi=10.1.1.296.7383. DOI: 10.1016/B978-0-12-384988-5.00037-1. 92

[CHHW94]   R. Calkin, R. Hempel, H.-C. Hoppe, and P. Wypior. 1994. Portable programming with the PARMACS message-passing library. *Parallel Computing*, 20(4): 615–632. DOI: 10.1016/0167-8191(94)90031-0. 63

[ChJR07]   Barbara Chapman, Gabriele Jost, and Ruud van der Pas. 2007. *Using OpenMP: Portable Shared Memory Parallel Programming (Scientific and Engineering Computation)*. The MIT Press. ISBN:0262533022 9780262533027. 64

[ChLi78]   S. K. Chang and C. N. Liu. 1978. Modeling and design of distributed information systems. In J. T. Tou (Ed.) *Advances in Information Systems Science*. Springer, Boston, MA, 157-221. 17

[ChLY18]   Xi Cheng, Xiang Li, and Jian Yang. 2018. SESR: Single image super resolution with recursive squeeze and excitation networks. *2018 24th International Conference on Pattern Recognition (ICPR)*, Beijing, 147–152. DOI: 10.1109/ICPR.2018.8546130. 89

[Clar17]   Lin Clark. 2017. Entering the Quantum Era—How Firefox got fast again and where it's going to get faster. Retrieved Dec. 15, 2019 from https://hacks.mozilla. org/2017/11/entering-the-quantum-era-how-firefox-got-fast-again-and-where-its-going-to-get-faster/. 131

[CoDe93]   G. W. Cook and E. J. Delp. 1993. The use of high performance computing in JPEG image compression. *Proceedings of 27th Asilomar Conference on Signals, Systems and Computers*, Pacific Grove, CA, 846–851 vol.2. DOI: 10.1109/ACSSC.1993.342450. 28

[Conw63]   Melvin E. Conway. 1963. A multiprocessor system design. *Proceedings of the November 12–14, 1963, Fall Joint Computer Conference (AFIPS '63 (Fall))*. ACM, New York, 139–146. DOI: 10.1145/1463822.1463838. 63

[Cray84]   Cray Research, Inc. 1984. *Cray Computer Systems Technical Note: Multitasking User Guide*. Retrieved Dec. 15, 2019 from https://archive.org/details/cray-multitasking. 64

[Croc97]   Thomas Crockett. 1997. An introduction to parallel rendering. *Parallel Computing*, 23(7): 819–843. DOI: 10.1016/S0167-8191(97)00028-8. 82

[Cutr18]   Ian Cutress. 2018. *Intel Expands 8th Gen Core: Core i9 on Mobile, Iris Plus, Desktop, Chipsets, and vPro*. Retrieved Dec 15, 2019 from https://www.anandtech.com/show/12607/intel-expands-8th-gen-core-core-i9-on-mobile-iris-plus-desktop-chipsets-and-vpro. 18

[DaCF03]   A. Darling, L. Carey, and W. Feng. 2003. The design, implementation, and evaluation of mpiBLAST. *4th International Conference on Linux Clusters: The HPC Revolution*. Retrieved Dec. 15, 2019 from https://public.lanl.gov/radiant/pubs/bio/cwce03.pdf. 84

[DaTo99]   F. Dahlgren and J. Torrellas. 1999. Cache-only memory architectures. *IEEE Computer*, 32(6): 72–79. DOI: 10.1109/2.769448. 33

[DCMC04]   Yuri Dotsenko, Cristian Coarfa, John Mellor-Crummey, and Daniel Chavarría-Miranda. 2004. Experiences with co-array fortran on hardware shared memory platforms. In Rudolf Eigenmann, Zhiyuan Li, and Samuel P. Midkiff (Eds.) *Proceedings of the 17th International Conference on Languages and Compilers for High Performance Computing*

(LCPC'04), Springer-Verlag, Berlin, Heidelberg, 332–347. DOI: 10.1007/11532378_24. 58, 63

[DeAr15]   Leonidas Deligiannidis and Hamd Arabnia. 2014. *Emerging Trends in Image Processing, Computer Vision and Pattern Recognition* (1st ed.). Morgan Kaufmann Publishers Inc., San Francisco, CA. 90

[Dela88]   H. C. Delany. 1988. Ray tracing on a connection machine. In J. Lenfant (Ed.) *Proceedings of the 2nd International Conference on Supercomputing (ICS '88)*, ACM, New York, 659–667. DOI: 10.1145/55364.55429. 83

[DeMi75]   Jack B. Dennis and David P. Misunas. 1975. A preliminary architecture for a basic data flow architecture. In *Proceedings of the 2nd Annual Symposium on Computer Architecture*. IEEE Press, New York, 126–132. DOI: 10.1145/641675.642111. 17, 23

[DGNP88]   F. Darema, D. A. George, V. A. Norton, and G. F. Pfister. 1988. A single-program-multiple-data computational model for EPEX/FORTRAN, *Parallel Computing*, 7(1): 11-24. DOI: 10.1016/0167-8191(88)90094-4. 61

[DoEl84]   Jack J. Dongarra and Stanley C. Eisenstat. 1984. Squeezing the most out of an algorithm in CRAY FORTRAN. *ACM Transactions on Mathematical Software*, 10(3): 219–230. DOI: 10.1145/1271.319413. 60

[Dong16]   Jack Dongarra. 2016. Report on the Sunway TaihuLight system. *University of Tennessee Department of Electrical Engineering and Computer Science Tech Report UT-EECS-16-742.* 46

[DoSt84]   J. J. Dongarra and G. W. Stewart. 1984. LINPACK—A package for solving linear systems. In W. R. Cowell (Ed.) *Sources and Development of Mathematical Software*, Prentice Hall, 20-48. 66

[DuSK85]   Augustin A. Dubrulle, Randolph G. Scarborough, and Harwood G. Kolsky. 1985. How to write a good vectorizable FORTRAN. IBM Palo Alto Scientific Center. *Technical Report No. G320-3478.* September. 60

[DWLT12]   Peng Du, Rick Weber, Piotr Luszczek, Stanimire Tomov, Gregory Peterson, and Jack Dongarra. 2012. From CUDA to OpenCL: Toward a performance-portable solution for multi-platform GPU programming. *Parallel Computing*, 38(8): 391–407. DOI: 10.1016/j.parco.2011.10.002. 135

[EBAS11]   H. Esmaeilzadeh, E. Blem, R. S. Amant, K. Sankaralingam and D. Burger. 2011. Dark silicon and the end of multicore scaling. 2011. *38th Annual International Symposium on Computer Architecture (ISCA)*, San Jose, CA, IEEE, 365–376. DOI: 10.1145/2024723.2000108. 107, 119

[ElCh01]   T. El-Ghazawi and S. Chauvin. 2001. UPC benchmarking issues. *International Conference on Parallel Processing*, Valencia, Spain, IEEE, 365–372. DOI: 10.1109/ICPP.2001.952082. 63

[ENSL18]   Ola Engkvist, Per-Ola Norrby, Nidhal Selmi, Yu-hong Lam, Zhengwei Peng, Edward Sherer, Willi Amberg, Thomas Erhard, and Lynette Smyth. 2018. Computational prediction of chemical reactions: Current status and outlook. *Drug Discovery Today*, 23. DOI: 10.1016/j.drudis.2018.02.014. 88

[EwMa78]   Robert H. Ewald and Lynn D. Maas. 1978. A high performance graphics system for the CRAY-1. In *Proceedings of the 5th Annual Conference on Computer Graphics and Interactive Techniques (SIGGRAPH '78)*. ACM, New York, 76–81. DOI: 10.1145/800248.807374. 71, 82

[FaRo15]   Hannes Fassold and Jakub Rosner. 2015. A real-time GPU implementation of the SIFT algorithm for large-scale video analysis tasks. In Nasser Kehtarnavaz and Matthias F. Carlsohn (Eds.) *Real-Time Image and Video Processing, International Society for Optics and Photonics*, SPIE, 9400, 62–69. DOI: 10.1117/12.2083201. 89

[FCCS92]   Federal Coordinating Council for Science, Engineering, and Technology Office of Science and Technology Policy. 1992. *Grand Challenges: High Performance Computing and Communications the FY1992 U.S. Research and Development Program*. Retrieved Dec. 15, 2019 from https://www.nitrd.gov/pubs/bluebooks/1992/pdf/bluebook92.pdf. 77, 133

[FERN84]   Joseph A. Fisher, John R. Ellis, John C. Ruttenberg, and Alexandru Nicolau. 1984. Parallel processing: a smart compiler and a dumb machine. In *Proceedings of the 1984 SIGPLAN Symposium on Compiler Construction (SIGPLAN '84)*. ACM, New York, 37–47. DOI: 10.1145/502874.502878. 32

[FiDo91]   J. R. Fischer and J. Dorband. 1991. Applications of the MasPar MP-1 at NASA/Goddard. *COMPCON Spring '91 Digest of Papers*, San Francisco, 278–282. DOI: 10.1109/CMPCON.1991.128818. 28

[FiRa81]   David Fincham and B. J. Ralston. 1981. Molecular dynamics simulation using the cray-1 vector processing computer. *Computer Physics Communications*, 23(2): 127-134. DOI: 10.1016/0010-4655(81)90027-8. 71, 86

[FiZe16]   Sergiy Yu. Fialko and Filip Zeglen. 2016. Preconditioned conjugate gradient method for solution of large finite element problems on CPU and GPU. *Journal of Telecommunication and Information Technology*, 2(1): 26–33. 80

[Flyn72]   Michael J. Flynn. 1972. Some computer organizations and their effectiveness. *IEEE Transactions on Computers*, 21(9): 948–960. DOI: 10.1109/TC.1972.5009071. 135

[FoSu05]   Tim Foley and Jeremy Sugerman. 2005. KD-tree acceleration structures for a GPU raytracer. In *Proceedings of the ACM SIGGRAPH/EUROGRAPHICS Conference on Graphics hardware (HWWS '05)*. ACM, New York, 15–22. DOI: 10.1145/1071866.1071869. 83

[FoxG18]   Geoffrey Fox. 2018. Private communication. 58, 62

[FQKY04]   Zhe Fan, Feng Qiu, Arie Kaufman, and Suzanne Yoakum-Stover. 2004. GPU cluster for high performance Computing. In *Proceedings of the 2004 ACM/IEEE Conference on Supercomputing (SC '04)*. IEEE Computer Society, Washington, DC, 47–47. DOI: 10.1109/SC.2004.26. 43

[Fraz17]   Bryant Frazer. July 2017. Five SIGGRAPH trends changing the way we make media. Retrieved Dec. 15, 2019 from http://www.studiodaily.com/2017/07/five-siggraph-trends-changing-way-make-media/. 83

[Fuch77]   Henry Fuchs. 1977. Distributing a visible surface algorithm over multiple processors. In *Proceedings of the 1977 Annual Conference (ACM '77)*. ACM, New York, 449–451. DOI: 10.1145/800179.810237. 71, 82

[Furt18]   M. Furtney. Private communication. 2018. 65

[GaKP81]   D.D. Gajski, D.J. Kuck, and D.A. Padua. 1981. Dependence driven computation. *Proceedings of COMPCON 81 Spring Computer Conference* (February 1981), 168–172. 66

[GaPS16]   Efstratios Gallopoulos, Bernard Philippe, and Ahmed H. Sameh. 2016. *Parallelism in Matrix Computations*. Springer, Scientific Computations Series. 76

[GiJo89] M. Ginsberg and J. P. Johnson. 1989. Benchmarking the performance of physical impact simulation software on vector and parallel computers, In J. L. Martin and S. F. Lundstrom, (Eds.) *Supercomputing 88: Volume 2—Science and Applications*, IEEE Computer Society Press, Washington, D.C. 180–190. 80, 97

[Gent82]   W. Gentzsch. 1982. Survey of new vector computers: The CRAY 1S from CRAY research; the CYBER 205 from CDC and the parallel computer from ICL—architecture and programming (January 1982). Retrieved Dec. 15, 2019 from https://ntrs.nasa.gov/archive/nasa/casi.ntrs.nasa.gov/19830026336.pdf. 57

[Gent84]   Wolfgang Gentzsch. 1984. Vectorization of a sample program, on different vector and parallel computers. In: *Vectorization of Computer Programs with Applications to Computational Fluid Dynamics*. Notes on Numerical Fluid Mechanics, 8. Vieweg+Teubner Verlag. DOI: 10.1007/978-3-322-87861-8_4. 59

[GlHo83]   Joseph Gloudeman and J. C. Hodge. 1983. The adaptation of MSC/NASTRAN to a supercomputer. In Manfred Ruschitzka (Ed.) *Parallel and Large-Scale Computers: Performance, Architecture, Applications, Proceedings of the IMACS World Congress on Systems*

Simulation and Scientific Computation, Montréal, Québec, Canada, 8–12 (August 1982), 185-189, North-Hollan. 71, 80

[Gins99]   Myron Ginsberg. 1999. Influences, challenges, and strategies for automotive HPC benchmarking and performance improvement. *Parallel Computing*, 25(12): 1459–1476. DOI: 10.1016/S0167-8191(99)00059-9. 81

[Gosd66]   J. A. Gosden. 1966. Explicit parallel processing description and control in programs for multi- and uni-processor computers. In *Proceedings of the November 7-10, 1966, Fall Joint Computer cConference (AFIPS '66 (Fall))*. ACM, New York, 651–660. DOI: 10.1145/1464291.1464361. 64

[GPKK82]   D. D. Gajski, D. A. Padua, D. J. Kuck, and R. H. Kuhn. 1982. A second opinion on data flow machines and languages. *IEEE Computer*, 15(2): 58–69. DOI: 10.1109/MC.1982.1653942. 23

[GuKW85]   J. R Gurd, C. C Kirkham, and I. Watson. 1985. The Manchester prototype data-flow computer. *Communications of ACM*, 28(1): 34–52. DOI: 10.1145/2465.2468. 23

[Gust88]   John L. Gustafson. 1988. Reevaluating Amdahl's law. *Communications of ACM*, 31(5): 532–533. DOI: 10.1145/42411.42415. 130

[HLJH09]   Jared Hoberock, Victor Lu, Yuntao Jia, and John C. Hart. 2009. Stream compaction for deferred shading. In *Proceedings of the Conference on High Performance Graphics 2009 (HPG '09)*, Stephen N. Spencer, David McAllister, Matt Pharr, and Ingo Wald (Eds.). ACM, New York, 173–180. DOI: 10.1145/1572769.1572797. 83

[HDDR16]   James Hegarty, Ross Daly, Zachary DeVito, Jonathan Ragan-Kelley, Mark Horowitz, and Pat Hanrahan. 2016. Rigel: flexible multi-rate image processing hardware. *ACM Transactions on Graphics*, 35(4): Article 85, 11 pages. DOI: 10.1145/2897824.2925892. 50

[HDYD12]   Geoffrey Hinton, Li Deng, Dong Yu, George E. Dahl, Abdel-rahman Mohamed, Navdeep Jaitly, Andrew Senior, Vincent Vanhoucke, Patrick Nguyen, Tara N. Sainath, and Brian Kingsbur. 2012. Deep neural networks for acoustic modeling in speech recognition. *IEEE Signal Processing Magazine*, 29(6): 82-97. DOI: 10.1109/MSP.2012.2205597. 92

[HeCh16]   James Heather and Benjamin Chain. 2016. The sequence of sequencers: The history of sequencing DNA. *Genomics*, 101(1): 1–8. DOI: 10.1016/j.ygeno.2015.11.003. 84

[Hend79]   Richard A. Hendrickson. 1979. Array processing extensions to FORTRAN. In *Proceedings of the 1979 Annual Conference (ACM '79)*, Arvid L. Martin and James L. Elshoff (Eds.). ACM, New York, 175–178. DOI: 10.1145/800177.810061. 57

[HePa19]   John l. Hennessey and David A. Patterson. 2019. A new golden age for computer architecture. *Communications of the ACM*, 62(2). DOI: 10.1145/3282307.103, 108, 119

[Hine18]   Jonathan Hines. Genomics code exceeds exaops on Summit supercomputer. 2018. https://www.olcf.ornl.gov/2018/06/08/genomics-code-exceeds-exaops-on-summit-supercomputer/. 104

[HiTa72]   R.G. Hintz, and D.P. Tats. 1972. Control data STAR-100 processor design. *COMPCON'72 Digest*, 1–4. 16

[Hoar78]   C. A. R. Hoare. 1978. Communicating sequential processes. *Communications of the ACM*, 21(8): 666–677. DOI: 10.1145/359576.359585. 62

[HPCA18]   Website of the HPC Advisory Council. http://hpcadvisorycouncil.com/. 77, 133

[HaAb11]   Tianyi David Han and Tarek S. Abdelrahman. 2011. Reducing branch divergence in GPU programs. *Proceedings of the Fourth Workshop on General Purpose Processing on Graphics Processing Units (GPGPU-4)*. ACM, New York, NY, Article 3, 8 pages. DOI: 10.1145/1964179.1964184. 61

[HuHL17]   Joseph Huber, Oscar Hernandez, and Graham Lopez. 2017. Effective vectorization with OpenMP 4.5. *Oak Ridge National Laboratory Technical Report ORNL/TM-2016/391*. 46, 47

[HuKa07]   Huang, C. and Kale, L. V. 2007. Charisma: Orchestrating migratable parallel objects. In *Proceedings of the 16th International Symposium on High Performance Distributed Computing 2007, HPDC'07*, 75–84. DOI: 10.1145/1272366.1272377. 125

[IRDS20a]   International Roadmap for Devices and System. 2020 Ed. More Moore. https://irds.ieee.org/. 117

[IRDS20b]   International Roadmap for Devices and System. 2020 Ed. Beyond CMOS. https://irds.ieee.org/. 117

[Iver62]   Kenneth E. Iverson. 1962. *A Programming Language*. John Wiley and Sons, Inc., New York. 56

[JKMM18]   William Jalby, David Kuck, Allen D. Malony, Michel Masella, Abdelhafi Mazouz and Mihail Popov. 2018. The long and winding road toward efficient high-performance computing. In *Proceedings of the IEEE*, 106(11): 1985–2003, DOI: 10.1109/JPROC.2018.2851190. 120, 133

[JLMA09]   Christopher Grant Jones, Rose Liu, Leo Meyerovich, Krste Asanović, and Rastislav Bodík. 2009. Parallelizing the web browser. In *Proceedings of the First USENIX Conference on Hot Topics in Parallelism (HotPar'09)*. USENIX Association, Berkeley, CA, 7–7. 131

[JuMa09]   Daniel Jurafsky and James Martin. 2009. Speech and Language processing. Pearson Prentice Hall, Upper Saddle River, NJ. 91

[JYPP17]   Norman P. Jouppi, Cliff Young, Nishant Patil, David Patterson, Gaurav Agrawal, Raminder Bajwa, Sarah Bates, Suresh Bhatia, Nan Boden, Al Borchers, Rick Boyle, Pierre-luc Cantin, Clifford Chao, Chris Clark, Jeremy Coriell, Mike Daley, Matt Dau, Jeffrey Dean, Ben Gelb, Tara Vazir Ghaemmaghami, Rajendra Gottipati, William Gulland, Robert Hagmann, C. Richard Ho, Doug Hogberg, John Hu, Robert Hundt, Dan Hurt, Julian Ibarz, Aaron Jaffey, Alek Jaworski, Alexander Kaplan, Harshit Khaitan, Daniel Killebrew, Andy Koch, Naveen Kumar, Steve Lacy, James Laudon, James Law, Diemthu Le, Chris Leary, Zhuyuan Liu, Kyle Lucke, Alan Lundin, Gordon MacKean, Adriana Maggiore, Maire Mahony, Kieran Miller, Rahul Nagarajan, Ravi Narayanaswami, Ray Ni, Kathy Nix, Thomas Norrie, Mark Omernick, Narayana Penukonda, Andy Phelps, Jonathan Ross, Matt Ross, Amir Salek, Emad Samadiani, Chris Severn, Gregory Sizikov, Matthew Snelham, Jed Souter, Dan Steinberg, Andy Swing, Mercedes Tan, Gregory Thorson, Bo Tian, Horia Toma, Erick Tuttle, Vijay Vasudevan, Richard Walter, Walter Wang, Eric Wilcox, and Doe Hyun Yoon. 2017. In-datacenter performance aAnalysis of a tensor processing unit. In *Proceedings of the 44th Annual International Symposium on Computer Architecture (ISCA '17)*. Association for Computing Machinery, New York, 1–12. DOI: 10.1145/3079856.3080246. 101, 104, 118, 120

[KaCo81]   Walter Karplus and Danny Cohen. 1981. Architectural and software issues in the design and application of peripheral array processors. *Computer*, 14(9): 11–17. DOI: 10.1109/C-M.1981.220594. 89

[KaHH04]   Laxmikant V. Kalé, Mark Hills, and Chao Huang. 2004. An orchestration language for parallel objects. In *Proceedings of the 7th Workshop on Workshop on Languages, Compilers, and Run-Time Support for Scalable Systems (LCR '04)*. ACM, New York, 1–6. DOI: 10.1145/1066650.1066658. 66

[KDLS86]   David J. Kuck, Edward S. Davidson, Duncan H. Lawrie, and Ahmed H. Sameh. 1986. Parallel supercomputing today and the Cedar approach. *Science*, 231(4741): 967–974. DOI: 10.1126/science.231.4741.967. 32

[KDSA08]   J. Kim, W. J. Dally, S. Scott, and D. Abts. 2008. Technology-driven, highly-scalable dragonfly topology. In *2008 International Symposium on Computer Architecture*, Beijing, 2008, 77–88, DOI: 10.1109/ISCA.2008.19. 45

[KeKZ11]   Ken Kennedy, Charles Koelbel, and Hans Zima. 2011. The rise and fall of high performance Fortran. *Communications of the ACM*, 54(11): 74–82. DOI: 10.1145/2018396.2018415. 57, 58

[KHAH16]   Hirak Kashyap, Hasin Ahmed, Nazrul Hoque, Swarup Roy, and Dhruba K. Bhattacharyya. 2016. Big data analytics in bioinformatics: Architectures, techniques, tools and issues. *Network Modeling Analysis in Health Informatics and Bioinformatics*, 5(Article 28): DOI: 10.1007/s13721-016-0135-4. 84, 85

[KiTG05]   J. Kim, W. J. Dally, B. Towles, and A. K. Gupta. 2005. Microarchitecture of a high radix router. *32nd International Symposium on Computer Architecture (ISCA'05)*, Madison, WI, 420–431. DOI: 10.1109/ISCA.2005.35. 45

[KMWG13]   David E Keyes, Lois C McInnes, Carol Woodward, William Gropp, Eric Myra, Michael Pernice, John Bell, Jed Brown, Alain Clo, Jeffrey Connors, Emil Constantinescu, Don Estep, Kate Evans, Charbel Farhat, Ammar Hakim, Glenn Hammond, Glen Hansen, Judith Hill, Tobin Isaac, Xiangmin Jiao, Kirk Jordan, Dinesh Kaushik, Efthimios Kaxiras, Alice Koniges, Kihwan Lee, Aaron Lott, Qiming Lu, John Magerlein, Reed Maxwell, Michael McCourt, Miriam Mehl, Roger Pawlowski, Amanda P Randles, Daniel Reynolds, Beatrice Rivière, Ulrich Rüde, Tim Scheibe, John Shadid, Brendan Sheehan, Mark Shephard, Andrew Siegel, Barry Smith, Xianzhu Tang, Cian Wilson, and Barbara Wohlmuth. 2013. Multiphysics simulations: Challenges and opportunities. *The International Journal of High Performance Computing Applications*, 27(1): 4–83. DOI: 10.1177/1094342012468181. 81

[KnWL93]   K. Knobe, M. Weiss, and J. D. Lukas. 1993. Optimization techniques for SIMD Fortran compilers. *Concurrency: Practice and Experience*, 5(7): 527–552. DOI: 10.1002/cpe.4330050703. 58

[Kris15]   Stig-Arne Kristoffersen. 2015. Future Trends of Seismic Analysis. Retrieved Dec. 17 from https://www.slideshare.net/StigArneKristoffersen/future-trends-of-seismic-analysis, 2015. 79

[KrSH12]   Alex Krizhevsky, Ilya Sutskever, and Geoffrey E. Hinton. 2017. ImageNet classification with deep convolutional neural networks. *Communications of the ACM*, 60(6): 84–90. DOI: 10.1145/3065386. 89

[Kuck18]   David Kuck. Private Communication. 2018. 64

[KuMC72]   D. J. Kuck, Y. Muraoka, and Shyh-Ching Chen. 1972. On the number of operations simultaneously executable in Fortran-like programs and their resulting speedup. *IEEE Transactions on Computers*, 21(12): 1293–1310. DOI: 10.1109/T-C.1972.223501. 59, 129

[KuNa17]   V. Kumar and A. Naeemi. 2017. An overview of 3D integrated circuits. *IEEE MTT-S International Conference on Numerical Electromagnetic and Multiphysics Model-*

ing and Optimization for RF, Microwave, and Terahertz Applications (NEMO), Seville, 311–313. DOI: https://doi.org/10.1109/NEMO.2017.7964270. 117

[KuPa81]   Robert H. Kuhn and David A. Padua (Eds.). 1981. Tutorial on parallel processing. Silver Spring, Md : IEEE Computer Society Press, c1981. Initially presented at the *Tenth International Conference on Parallel Processing*, August 25–28, 1981, Bellaire, Michigan. 1

[KuYB16]   Gregorij Kurillo, Allen Y. Yang, and Ruzena Bajcsy. 2016. *3D Telepresence for Reducing Transportation Costs*. EECS Department, University of California, Berkeley. Nov., 2016. UCB/EECS-2016-168. Retrieved Dec. 17, 2019 from https://www2.eecs.berkeley.edu/Pubs/TechRpts/2016/EECS-2016-168.pdf. 83

[Ledf10]   Heidi Ledford. 2010. Supercomputer sets protein folding record. *Nature*, Oct. 2010. DOI: 10.1038/news.2010.541. 87

[Leas18]   Bruce Leasure. Private Communication. 2018. 64

[Leis85]   Charles E. Leiserson. 1985. Fat-trees: Universal networks for hardware-efficient supercomputing. *IEEE Transactions on Computers*, C-34(10): 892–901. DOI: 10.1109/TC.1985.6312192. 45

[LeSB88]   Thomas J. LeBlanc, Michael L. Scott, and Christopher M. Brown. 1988. Large-scale parallel programming: experience with BBN butterfly parallel processor. *Proceedings of the ACM/SIGPLAN Conference on Parallel Programming: Experience with Applications, Languages, and Systems (PPEALS '88)*, Richard L. Wexelblat (Ed.). ACM, New York, 161–172. DOI: 10.1145/62115.62131. 37

[LFKL93]   P. Geoffrey Lowney, Stefan M. Freudenberger, Thomas J. Karzes, W. D. Lichtenstein, Robert P. Nix, John S. O'Donnell, and John Ruttenberg. 1993. The multiflow trace scheduling compiler. *Journal of Supercomputing*, 7(1–2): 51–142. DOI: 10.1007/BF01205182. 59

[LHKK79]   C. L. Lawson, R. J. Hanson, D. R. Kincaid, and F. T. Krogh. 1979. Basic linear algebra subprograms for Fortran usage. *ACM Transactions on Mathematical Software*. 5(3): 308–323. DOI: 10.1145/355841.355847. 80

[LLBR75]   D. H. Lawrie, T. Layman, D. Baer, and J. M. Randal. 1975. Glypnir—a programming language for Illiac IV. *Communications of the ACM*, 18(3): 157–164. DOI: 10.1145/360680.360687. 61

[LLNL02]   Lawrence Livermore National Lab. From Kilobytes to Petabytes in 50 Years. March 2002. Retrieved Dec. 17, 2019, currently found at https://www.eurekalert.org/features/doe/2002-03/dlnl-fkt062102.php. 133

[Lori72]   Harold Lorin. 1972. *Parallelism in Hardware and Software: Real and Apparent Concurrency.* Prentice-Hall. Jan. 1, 1972. 62

[Lowe04]   David G Lowe. 2004. Distinctive image features from scale-invariant key-points. *International Journal of Computer Vision*, 60(2): 91–110. DOI: 10.1023/B:V ISI.0000029664.99615.94. 89

[LTEK20]   Charles E. Leiserson, Neil C. Thompson, Joel S. Emer, Bradley C. Kuszmaul, Butler W. Lampson, Daniel Sanchez, and Tao B. Schardl. 2020. There's plenty of room at the Top: What will drive computer performance after Moore's law? *Science*, 368(6495): DOI: 10.1126/science.aam9744. 72, 104, 120

[LTZG17]   Shaoshan Liu, Jie Tang, Zhe Zhang, and Jean-Luc Gaudiot. 2017. CAAD: Computer Architecture for Autonomous Driving. https://arxiv.org/pdf/1702.01894. 90

[LuBa86]   S. F. Lundstrom and G. H. Barnes. 1986. A controllable MIMD architecture. In Dharma P Agrawal (Ed.) *Advanced Computer Architecture*, IEEE Computer Society Press, Los Alamitos, CA, 30–38. 16, 64

[LZHS18]   Shih-Chieh Lin, Yunqi Zhang, Chang-Hong Hsu, Matt Skach, Md E. Haque, Lingjia Tang, and Jason Mars. 2018. The architectural implications of autonomous driving: Constraints and acceleration. *SIGPLAN* Not. 53(2): 751–766. DOI: 10.1145/3296957.3173191. 90

[MAKD17]   V. Mironov, Y. Alexeev, K. Keipert, M. D'mello, A. Moskovsky, and M. S. Gordan. An efficient MPI/OpenMP parallelization of the Hartree-Fock method for the second generation of Intel® Xeon PhiTM processor. S*uperComputing* 17. 87

[Mali19]   Jitendra Malik. 2019. Deep learning in computer vision. *National Academy of Sciences Arthur M. Sackler Colloquium: The Science of Deep Learning.* Washington, D.C. March 13–14, 2019. https://www.youtube.com/watch?v=PHbaaqgYKwU. 90

[MasP92a]   *MasPar Fortran Reference Manual*, Document Part Number 9302-0001, Revision A3, July 1992. Software Version 2.0 MasPar Computer Corporation, Sunnyvale CA. 56

[MasP92b]   *MasPar Programming Language (ANSI C compatible MPL) Reference Manual.* Software Version 3.0. Document Part Number: 9302-0001. Revision: A3 July 1992. MasPar Computer Corporation. 61

[McCo63]   B. H. McCormick. 1963. The Illinois pattern recognition computer-ILLIAC III. *IEEE Transactions on Electronic Computers*, EC-12(6): 791–813. DOI: 10.1109/ PGEC.1963.263562. 71, 88

[MCCS16]   Timothy G. Mattson, Romain Cledat, Vincent Cavé, Vivek Sarkar, Zoran Budimlic, Sanjay Chatterjee, Joshua B. Fryman, Ivan Ganev, Robin Knauerhase, Min Lee,

Benoît Meister, Brian Nickerson, Nick Pepperling, Bala Seshasayee, Sagnak Tasirlar, Justin Teller, and Nick Vrvilo. 2016. The open community runtime: A runtime system for extreme scale computing. *IEEE High Performance Extreme Computing Conference (HPEC)*, Waltham, MA, 1–7. DOI: 10.1109/HPEC.2016.7761580. 66

[McRK85]   Robert N. Mcdonough, Barry E. Raff, and Joyce L. Kerr. 1985. Image formation from spaceborne synthetic aperture radar signals. *Johns Hopkins APL Technical Digest*, 6(4): 300–312. Retrieved Dec. 17, 2019 from https://www.jhuapl.edu/Content/techdigest/pdf/V06-N04/06-04-McDonough.pdf. 89

[MGGW11]   Saeed Maleki, Yaoqing Gao, Maria J. Garzarán, Tommy Wong, and David A. Padua. 2011. An evaluation of vectorizing compilers. In *Proceedings of the 2011 International Conference on Parallel Architectures and Compilation Techniques (PACT '11)*. IEEE Computer Society, Washington, DC, 372–382. DOI: 10.1109/PACT.2011.68. 60

[Mill73]   Robert E. Millstein. 1973. Control structures in ILLIAC IV Fortran. *Communications of the ACM*, 16(10): 621–627. DOI: 10.1145/362375.362398. 56

[Mitt18]   Sparsh Mittal. 2018. A survey of FPGA-based accelerators for convolutional neural networks. *Neural Computing and Applications*, 32, 1109–1139 DOI: 10.1007/s00521-018-3761-1. 50

[MiVe15]   Sparsh Mittal and Jeffery S. Vetter. 2015. A survey of CPU-GPU heterogeneous computing techniques. *ACM Computing Surveys*, 47(4): 69:1–69:35. DOI: 10.1145/2788396. 43

[MLBV17]   Mike Mikailov, Fu-Jyh Luo, Stuart Barkley, Lohit Valleru, Stephen Whitney, Zhichao Liu, Shraddha Thakkar, Weida Tong, and Nicholas Petrick. (2017). Scaling bioinformatics applications on HPC. *BMC Bioinformatics*. DOI: 10.1186/s12859-017-1902-7. 85

[Moor18]   Samuel K. Moore. 2018. Researchers Invent a Way to Speed Intel's 3D XPoint Computer Memory. Retrieved Dec. 17, 2019 from https://spectrum.ieee.org/tech-talk/semiconductors/memory/engineers-invent-a-way-to-speed-intels-3d-xpoint-computer-memory. 116

[MPIF93]   The MPI Forum. 1993. MPI: a message passing interface. In *Proceedings of the 1993 ACM/IEEE Conference on Supercomputing (Supercomputing '93)*. ACM, New York, 878–883. DOI: 10.1145/169627.169855. 63

[MSCC13]   Tomas Mikolov, Ilya Sutskever, Kai Chen, Greg Corrado, and Jeffrey Dean. 2013. Distributed representations of words and phrases and their compositionality. In *Proceedings of the 26th International Conference on Neural Information Processing Systems - Volume 2*

(NIPS'13), C. J. C. Burges, L. Bottou, M. Welling, Z. Ghahramani, and K. Q. Weinberger (Eds.), Vol. 2. Curran Associates Inc., 3111–3119. 92

[MSGS18]   Adam Marrs, Josef Spjut, Holger Gruen, Rahul Sathe, and Morgan McGuire. 2018. Adaptive temporal antialiasing. In *Proceedings of the Conference on High-Performance Graphics (HPG '18)*. ACM, New York, Article 1, 4 pages. DOI: 10.1145/3231578.3231579. 83

[Muus87]   Michael John Muuss. 1987. RT and REMRT: Shared memory parallel and network distributed ray-tracing programs. *Proceedings of 4th Computer Graphics Workshop*, Cambridge, MA, Usenix Association, 86–98. 83

[Nick90]   J. R. Nickolls. 1990. The design of the MasPar MP-1: a cost effective massively parallel computer. *Digest of Papers Compcon Spring '90*. Thirty-Fifth IEEE Computer Society International Conference on Intellectual Leverage, San Francisco, CA, 25–28. DOI: 10.1109/CMPCON.1990.63649. 28

[NNIK92]   K. Nakazawa, H. Nakamura, H. Imori, and S. Kawabe. 1992. Pseudo vector processor based on register-windowed superscalar pipeline. In *Supercomputing '92: Proceedings of the 1992 ACM/IEEE Conference on Supercomputing*, Minneapolis, MN, 642–651. DOI: 10.1109/SUPERC.1992.236638. 21

[Numr11]   Robert W. Numrich. 2011. Coarray Fortran. In David Padua (ed.), *Encyclopedia of Parallel Computing*. Springer Publishing Company, Incorporated, 304–310. 58, 63

[Nvid18]   Nvidia RTX Ray Tracing. Retrieved Dec. 17, 2019 from https://developer.nvidia.com/rtx/raytracing. 83

[OLGH07]   John D. Owens, David Luebke, Naga Govindaraju, Mark Harris, Jens Krüger, Aaron E. Lefohn, and Tim Purcell. 2007. A survey of general-purpose computation on graphics hardware. *Computer Graphics Forum*, 26(1): 80–113. DOI: 10.1111/j.1467-8659.2007.01012.x. 39

[PaCF91]   The Parallel Computing Forum. 1991. PCF parallel Fortran extensions. *SIGPLAN Fortran Forum*, 10(3): 1–57. DOI: 10.1145/122391.122392. 64

[Padu11]   Padua David (Ed.). 2011. *Encyclopedia of Parallel Computing*. Springer, Boston, MA. 134

[Park80]   D. S. Parker, Jr. 1980. Notes on shuffle/exchange-type switching networks. *IEEE Transactions on Computing*, 29(3): 213–222. DOI: 10.1109/TC.1980.1675553. 44

[PaWi75]   G. Paul and M. W. Wilson. 1975. The VECTRAN language: An experimental language for vector/matrix array processing. IBM Palo Alto Scientific Center. *Report G320-3334*, August. 57

[PeSc17]   J. R. Perilla and K. Schulten. Physical properties of the HIV-1 capsid from all-atom molecular dynamics simulations. *Nature Communications*. 8(15959). DOI: 10.1038/ncomms15959. 86

[PhFe05]   Matt Pharr and Randima Fernando. 2005. *GPU Gems 2: Programming Techniques for High-Performance Graphics and General-Purpose Computation (Gpu Gems)*, Part IV. Addison-Wesley Professional. ISBN:0321335597. 39

[Pier94]   Paul Pierce. 1994. The NX message passing interface. *Parallel Computing*, 20(4): 463–480. DOI: 10.1016/0167-8191(94)90023-X. 62

[PoIT18]   Mariya Popova, Olexandr Isayev, and Alexander Tropsha. 2018. Deep reinforcement learning for de novo drug design. *Science Advances*, [s.l.], 4(7): 1–14. DOI: 10.1126/sciadv.aap7885. 88

[Poly87]   C. D. Polychronopoulos and D. J. Kuck. 1987. Guided self-scheduling: A practical scheduling scheme for parallel supercomputers. *IEEE Transactions on Computing*, 36(12): 1425–1439. DOI: 10.1109/TC.1987.5009495. 65

[PrJo75]   David L. Presberg. 1975. The paralyzer: Ivtran's parallelism analyzer and synthesizer. In *Proceedings of the Conference on Programming Languages and Compilers for Parallel and Vector Machines*. ACM, New York, 9–16. DOI: 10.1145/800026.808396. 59

[PZKK02]   J.C. Phillips, G. Zheng, S. Kumar, and L.V. Kale. 2002. NAMD: Biomolecular simulation on thousands of processors. *Proceedings of the 2002 ACM/IEEE Conference on Supercomputing*, Baltimore, MD, 1–18. DOI: 10.1109/SC.2002.10019. 86

[Rahm64]   Aneesur Rahmen. 1964. Correlations in the motion of atoms in liquid argon. *Physical Review Journals Archive*, 136(2A): 405–411. DOI: 10.1103/PhysRev.136.A405. 86

[RauR94]   B. Ramakrishna Rau. 1994. Iterative modulo scheduling: an algorithm for software pipelining loops. *Proceedings of the 27th Annual International Symposium on Microarchitecture (MICRO 27)*. ACM, New York, 63–74. DOI: 10.1145/192724.192731. 59

[RDSK15]   Olga Russakovsky, Jia Deng, Hao Su, Jonathan Krause, Sanjeev Satheesh, Sean Ma, Zhiheng Huang, Andrej Karpathy, Aditya Khosla, Michael Bernstein, Alexander C. Berg, and Li Fei-Fei. 2015. ImageNet large scale visual recognition challenge. *International Journal of Computer Vision*, 115(3): 211–252. DOI: 10.1007/s11263-015-0816-y. 89

[Redd73]   S. F. Reddaway. 1973. DAP—a distributed array processor. In *Proceedings of the 1st Annual Symposium on Computer Architecture (ISCA '73)*, S. A. Szygenda (Ed.). ACM, New York, N 61–65. DOI: 10.1145/800123.803971. 17

[ReEd89]   Mosche Reshef and Mickey Edwards. Three dimensional seismic processing, migration, and modeling using parallel processing on Cray supercomputers. *Handbook of Geophysical Exploration: Seismic Exploration*, Elsevier, 21(Chapter 2): 11–34. DOI: 10.1016/B978-0-08-037018-7.50006-4. 78

[Reev84]   Anthony P. Reeves. 1984. Parallel Pascal: An extended Pascal for parallel computers. *Journal of Parallel and Distributed Computing*, 1(1): 64–80. DOI: 10.1016/0743-7315(84)90011-X. 56

[Rein07]   James Reinders. 2007. *Intel Threading Building Blocks* (1st ed.). O'Reilly and Associates, Inc., Sebastopol, CA. 65

[RHSD17]   Cosmin Radoi, Stephan Herhut, Jaswanth Sreeram, and Danny Dig. 2015. Are web applications ready for parallelism?. In *Proceedings of the 20th ACM SIGPLAN Symposium on Principles and Practice of Parallel Programming (PPoPP 2015)*. ACM, New York, 289–290. DOI: 10.1145/2688500.2700995. 131

[RoSe00]   T. Rognes and E. Seeberg. 2000 Six-fold speed-up of Smith-Waterman sequence database searches using parallel processing on common microprocessors. *Bioinformatics*, 16(8): 699–706. DOI: 10.1093/bioinformatics/16.8.699. 84

[Russ78]   Richard M. Russell. 1978. The CRAY-1 computer system. *Communications of the ACM*, 21(1): 63–72. DOI: 10.1145/359327.359336. 16

[Sala17]   Ryan Salazar. 2017. PIXAR Talks Render Farms with Ryan Salazar. Retrieved Dec. 17, 2019 from http://www.broadcastbeat.com/pixar-talks-render-farms-ryan-salazar/. 83

[SCBM98]   Allan Snavely, Larry Carter, Jay Boisseau, Amit Majumdar, Kang Su Gatlin, Nick Mitchell, John Feo, and Brian Koblenz. 1998. Multi-processor performance on the Tera MTA. In *Proceedings of the 1998 ACM/IEEE Conference on Supercomputing (SC '98)*. IEEE Computer Society, Washington, DC, 1–8. DOI: 10.1109/SC.1998.10049. 33

[SCOF98]   Carlos Sosa, John Carpenter, J. Octerski, and Michael Frisch. 1998. Ab initio quantum chemistry on the Cray T3E massively parallel supercomputer: II. *Journal of Computational Chemistry*, 19(9): 1053–1063. DOI: 10.1002/(SICI)1096-987X(19980715)19:9<1053::AID-JCC6>3.0.CO;2-P. 87

[SaHS09]   Vivek Sarkar, William Harrod, and Allan Snavely. 2009. Software challenges in extreme scale systems. *Journal of Physics: Conference Series*, 180, 1. DOI: 10.1088/1742-6596/180/1/012045. 41

[SBLO77]   Richard J. Swan, Andy Bechtolsheim, Kwok-Woon Lai, and John K. Ousterhout. 1977. The implementation of the Cm* multi-microprocessor. In *Proceedings of the June*

13-16, 1977, National Computer Conference (AFIPS '77). ACM, New York, 645–655. DOI: 10.1145/1499402.1499516. 16

[SCHL03]   Alex Settle, Daniel A. Connors, Gerolf Hoflehner, and Dan Lavery. 2003. Optimization for the Intel® Itanium® architecture register stack. In *Proceedings of the International Symposium on Code Generation and Optimization: Feedback-Directed and Runtime Optimization (CGO '03).* IEEE Computer Society, Washington, DC, 115–124. DOI: 10.1109/CGO.2003.1191538. 59

[Seit85]   Charles L. Seitz. 1985. The cosmic cube. *Communications of the ACM,* 28(1): 22–33. DOI: 10.1145/2465.2467. 62

[SJGH93]   J. P. Singh, T. Joe, J. L. Hennessy, and A. Gupta. 1993. An empirical comparison of the Kendall Square Research KSR-1 and Stanford DASH multiprocessors. In *Proceedings of the 1993 ACM/IEEE Conference on Supercomputing (Supercomputing '93).* ACM, New York, 214–225. DOI: 10.1145/169627.169699. 33

[SGMA10]   John A. Stratton, Vinod Grover, Jaydeep Marathe, Bastiaan Aarts, Mike Murphy, Ziang Hu, and Wen-mei W. Hwu. 2010. Efficient compilation of fine-grained SPMD-threaded programs for multicore CPUs. In *Proceedings of the 8th Annual IEEE/ACM International Symposium on Code Generation and Optimization,* pp. 111–119. ACM, DOI: 10.1145/1772954.1772971. 65

[Shal20]   John Shalf. 2020. The future of computing beyond Moore's Law, *Philosophical Transactions of the Royal Society A,* 378(2166). DOI: 10.1098/rsta.2019.0061. 117, 118

[Shaw09]   David E. Shaw, Ron O. Dror, John K. Salmon, J. P. Grossman, Kenneth M. Mackenzie, Joseph A. Bank, Cliff Young, Martin M. Deneroff, Brannon Batson, Kevin J. Bowers, Edmond Chow, Michael P. Eastwood, Douglas J. Ierardi, John L. Klepeis, Jeffrey S. Kuskin, Richard H. Larson, Kresten Lindorff-Larsen, Paul Maragakis, Mark A. Moraes, Stefano Piana, Yibing Shan, and Brian Towles. 2009. Millisecond-scale molecular dynamics simulations on Anton. In *Proceedings of the Conference on High Performance Computing Networking, Storage and Analysis (SC '09).* ACM, New York, Article 39, 11 pages. DOI: 10.1145/1654059.1654099. 87

[Shoo80]   Rita Shoor. 19080. CDC 205 runs 800 million operation/sec. *Computerworld,* 14(23): 1–2. 16

[SHPS16]   John E. Stone, Antti Pekka Hynninen, James C. Phillips, and Klaus J Schulten. 2016. Early experiences porting NAMD and VMD molecular simulation and analysis software to GPU-accelerated OpenPOWER platforms. *High Performance Computing. ISC High Performance 2016 International Workshops, ExaComm, E-MuCoCoS, HPC-*

IODC, IXPUG, IWOPH, P^3MA, VHPC, WOPSSS, Frankfurt, Germany, June 19–23, 2016, Revised Selected Papers. Springer International Publishing, 188–206. 62, 86

[SLJS15]   Christian Szegedy, Wei Liu, Yangqing Jia, Pierre Sermanet, Scott Reed, Dragomir Anguelov, Dumitru Erhan, Vincent Vanhoucke, and Andrew Rabinovich. 2015. Going deeper with convolutions, *IEEE Conference on Computer Vision and Pattern Recognition (CVPR)*, Boston, MA, 1–9. DOI: 10.1109/CVPR.2015.7298594. 89

[Smit78]   B. J. Smith. 1986. A pipelined, shared resource MIMD computer. In Dharma P Agrawal (Ed.), *Advanced Computer Architecture*, IEEE Computer Society Press, Los Alamitos, CA, 39–41. 17, 32

[Smit82]   Burton J. Smith. 1982. Architecture and applications of the HEP multiprocessor computer system. *Proceedings SPIE 0298, Real-Time Signal Processing IV*, (July 30, 1982). DOI: 10.1117/12.932535. 65

[Snir13]   Marc Snir. Supercomputing: The Coming Decade. Retrieved December 18, 2019 from http://extremecomputingtraining.anl.gov/files/2014/01/dinner-talk-8-13.pdf. 41

[SSBD95]   Thomas L. Sterling, Daniel Savarese, Donald J. Becker, John E. Dorband, Udaya A. Ranawake, and Charles V. Packer. 1995. BEOWULF: A parallel workstation for scientific computation. International Conference on Parallel Processing (ICPP), 11–14. 39, 51

[Stad79]   R. Staden. 1979. A strategy for DNA sequencing employing computer programs. *Nucleic Acid Research*, 6(7): 2601–2610. DOI: 10.1093/nar/6.7.2601. 84

[Stan16]   Stanford 100 Year Study of AI Panel. Artificial Intelligence in Life 2030. 2016. Retrieved Dec. 18, 2019 from https://ai100.stanford.edu/sites/default/files/ai_100_report_0906fnlc_single.pdf. 92

[StDo04]   Aad J. Van der Steen and Jack J. Dongarra. Overview of Recent Supercomputers. 2004. Retrieved Dec. 18, 2019 from http://www.netlib.org/utk/papers/advanced-computers/overview.html. 134

[Ster99]   Thomas Lawrence Sterling. (Eds.) *How to Build a Beowulf: A Guide to the Implementation and Application of PC Clusters*. MIT Press. Cambridge, MA. 39

[Stok77]   Richard A. Stokes. 1977. Burroughs scientific processor. In David J. Kuck, Duncan H. Lawrie, and Ahmed Sameh (Eds.) *Proceedings of a Symposium on High Speed Computer and Algorithm Organization*, held in Champaign, IL, April 13–15, 1977, Academic Press, New York. 16

[Ston77]   H. S. Stone. 1977. Multiprocessor scheduling with the aid of network flow algorithms. *IEEE Transactions in Software Engineering*, 3(1): 85–93. DOI: 10.1109/TSE.1977.233840. 17

[Sund90]   V. S. Sunderam. 1990. PVM: a framework for parallel distributed computing. *Concurrency: Practice and Experience*, 2(4): 315–339. DOI: 10.1002/cpe.4330020404. 63

[SVYM91]   Qasim Sheikh, Phuong Vu, Chao Yang, and Michael Merchant. 1992. Implementation of the Level 2 and 3 BLAS on the CRAY Y-MP and the CRAY-2. *Journal of Supercomputing*, 5(4): 291–305. DOI: 10.1007/BF00127950. 66

[SwFS77]   R. J. Swan, S. H. Fuller, and D. P. Siewiorek. 1977. Cm*: a modular, multi-microprocessor. In *Proceedings of the June 13-16, 1977, National Computer Conference (AFIPS '77)*. ACM, New York, 637–644. DOI: 10.1145/1499402.1499515. 16

[SwIB13]   Swiss Institute of Bioinformatics. 2013. Directory of computer-aided Drug Design tools. Retrieved Dec. 18, 2019 from https://www. click2drug.org. 98

[TAAD13]   Josep Torrellas, Sarita V. Adve, Vikram S. Adve, Danny Dig, Minh N. Do, Maria Jesus Garzaran, John C. Hart, Thomas S. Huang, Wen-mei W. Hwu, Samuel T. King, Darko Marinov, Klara Nahrstedt, David A. Padua, Madhusudan Parthasarathy, Sanjay J. Patel, and Marc Snir. 2013. Making parallel programming easy: Research contributions from Illinois. Illinois Parallelism Center. *IDEALS*: http://hdl.handle.net/2142/101902. 130

[TBRT02]   J. Tuck, L. Baugh, J. Renu, and J. Torellas. 2002. Sphinx Parallelization. Retrieved December 18, 2019 from https://www.ideals.illinois.edu/bitstream/handle/2142/11073/Sphinx%20Parallelization.pdf?sequence=2. 92

[THBC18]   S. Taheri, J. Heo, P. Behnam, J. Chen, A. Veidenbaum, and A. Nicolau. 2018. Acceleration framework for FPGA implementation of openVX graph pipelines. *IEEE 26th Annual International Symposium on Field-Programmable Custom Computing Machines (FCCM)*, Boulder, CO, 227–227. DOI: 10.1109/FCCM.2018.00061. 50

[Thin91]   *CM Fortran Programmer Guide*. Version 1.0. 1991. Thinking Machines Corporation, Cambridge, MA. January. 56

[TeMS87]   Jack A. Test, Mat Myszewski, and Richard C. Swift. 1987. The Alliant FX/Series: A language driven architecture for parallel processing of Dusty Deck Fortran. In J. W. de Bakker, A. J. Nijman, and Philip C. Treleaven (Eds.), *Proceedings of the Parallel Architectures and Languages Europe, Volume I: Parallel Architectures PARLE*, Springer-Verlag, London, UK, 345–356. 59

[Thor70]   J. E. Thornton. 1970. *Design of a Computer the Control Data 6600*. Scott Foresman and Company, Glenview Illinois, USA, 1970. 33

[ThSp18]   Neil Thompson and Svenja Spanuth. 2018. The Decline of Computers As a General Purpose Technology: Why Deep Learning and the End of Moore's Law are Fragmenting Computing. (Nov. 2018). Available at SSRN: http://dx.doi.org/10.2139/ssrn.3287769. 103, 108, 116, 117, 120

[TrOl10]   O. Trott and A. J. Olson. 2010. AutoDock Vina: Improving the speed and accuracy of docking with a new scoring function, efficient optimization, and multithreading. *Journal of Computational Chemistry*, 31(2): 455–461. DOI: 10.1002/jcc.21334. 88

[TOP500]   The TOP500 list. https://www.top500.org/. 2, 133

[TuRo88]   L. W. Tucker and G. G. Robertson. 1988. Architecture and applications of the Connection Machine. *Computer*, 21(8): 26–38. DOI: 10.1109/2.74. 28

[Unger58]   S. H. Unger. 1958. A computer oriented toward spatial problems. In *Proceedings of the May 6-8, 1958, Western Joint Computer Conference: Contrasts in Computers (IRE-ACM-AIEE '58 (Western))*. ACM, New York, 234–239. DOI: 10.1145/1457769.1457836. 88

[Vent01] Craig Venter, Mark D. Adams, Eugene W. Myers, Peter W. Li, Richard J. Mural, Granger G. Sutton, Hamilton O. Smith, Mark Yandell, Cheryl A. Evans, Robert A. Holt, Jeannine D. Gocayne, Peter Amanatides, Richard M. Ballew, Daniel H. Huson, Jennifer Russo Wortman, Qing Zhang, Chinnappa D. Kodira, Xiangqun H. Zheng, Lin Chen, Marian Skupski, Gangadharan Subramanian, Paul D. Thomas, Jinghui Zhang, George L. Gabor Miklos, Catherine Nelson, Samuel Broder, Andrew G. Clark, Joe Nadeau, Victor A. McKusick, Norton Zinder, Arnold J. Levine, Richard J. Roberts, Mel Simon, Carolyn Slayman, Michael Hunkapiller, Randall Bolanos, Arthur Delcher, Ian Dew, Daniel Fasulo, Michael Flanigan, Liliana Florea, Aaron Halpern, Sridhar Hannenhalli, Saul Kravitz, Samuel Levy, Clark Mobarry, Knut Reinert, Karin Remington, Jane Abu-Threideh, Ellen Beasley, Kendra Biddick, Vivien Bonazzi, Rhonda Brandon, Michele Cargill, Ishwar Chandramouliswaran, Rosane Charlab, Kabir Chaturvedi, Zuoming Deng, Valentina Di Francesco, Patrick Dunn, Karen Eilbeck, Carlos Evangelista, Andrei E. Gabrielian, Weiniu Gan, Wangmao Ge, Fangcheng Gong, Zhiping Gu, Ping Guan, Thomas J. Heiman, Maureen E. Higgins, Rui-Ru Ji, Zhaoxi Ke, Karen A. Ketchum, Zhongwu Lai, Yiding Lei, Zhenya Li, Jiayin Li, Yong Liang, Xiaoying Lin, Fu Lu, Gennady V. Merkulov, Natalia Milshina, Helen M. Moore, Ashwinikumar K Naik, Vaibhav A. Narayan, Beena Neelam, Deborah Nusskern, Douglas B. Rusch, Steven Salzberg, Wei Shao, Bixiong Shue, Jingtao Sun, Zhen Yuan Wang, Aihui Wang, Xin Wang, Jian Wang, Ming-Hui Wei, Ron Wides, Chunlin Xiao, Chunhua Yan, Alison Yao, Jane Ye, Ming Zhan, Weiqing Zhang,

Hongyu Zhang, Qi Zhao, Liansheng Zheng, Fei Zhong, Wenyan Zhong, Shiaoping C. Zhu, Shaying Zhao, Dennis Gilbert, Suzanna Baumhueter, Gene Spier, Christine Carter, Anibal Cravchik, Trevor Woodage, Feroze Ali, Huijin An, Aderonke Awe, Danita Baldwin, Holly Baden, Mary Barnstead, Ian Barrow, Karen Beeson, Dana Busam, Amy Carver, Angela Center, Ming Lai Cheng, Liz Curry, Steve Danaher, Lionel Davenport, Raymond Desilets, Susanne Dietz, Kristina Dodson, Lisa Doup, Steven Ferriera, Neha Garg, Andres Gluecksmann, Brit Hart, Jason Haynes, Charles Haynes, Cheryl Heiner, Suzanne Hladun, Damon Hostin, Jarrett Houck, Timothy Howland, Chinyere Ibegwam, Jeffery Johnson, Francis Kalush, Lesley Kline, Shashi Koduru, Amy Love, Felecia Mann, David May, Steven McCawley, Tina McIntosh, Ivy McMullen, Mee Moy, Linda Moy, Brian Murphy, Keith Nelson, Cynthia Pfannkoch, Eric Pratts, Vinita Puri, Hina Qureshi, Matthew Reardon, Robert Rodriguez, Yu-Hui Rogers, Deanna Romblad, Bob Ruhfel, Richard Scott, Cynthia Sitter, Michelle Smallwood, Erin Stewart, Renee Strong, Ellen Suh, Reginald Thomas, Ni Ni Tint, Sukyee Tse, Claire Vech, Gary Wang, Jeremy Wetter, Sherita Williams, Monica Williams, Sandra Windsor, Emily Winn-Deen, Keriellen Wolfe, Jayshree Zaveri, Karena Zaveri, Josep F. Abril, Roderic Guigó, Michael J. Campbell, Kimmen V. Sjolander, Brian Karlak, Anish Kejariwal, Huaiyu Mi, Betty Lazareva, Thomas Hatton, Apurva Narechania, Karen Diemer, Anushya Muruganujan, Nan Guo, Shinji Sato, Vineet Bafna, Sorin Istrail, Ross Lippert, Russell Schwartz, Brian Walenz, Shibu Yooseph, David Allen, Anand Basu, James Baxendale, Louis Blick, Marcelo Caminha, John Carnes-Stine, Parris Caulk, Yen-Hui Chiang, My Coyne, Carl Dahlke, Anne Deslattes Mays, Maria Dombroski, Michael Donnelly, Dale Ely, Shiva Esparham, Carl Fosler, Harold Gire, Stephen Glanowski, Kenneth Glasser, Anna Glodek, Mark Gorokhov, Ken Graham, Barry Gropman, Michael Harris, Jeremy Heil, Scott Henderson, Jeffrey Hoover, Donald Jennings, Catherine Jordan, James Jordan, John Kasha, Leonid Kagan, Cheryl Kraft, Alexander Levitsky, Mark Lewis, Xiangjun Liu, John Lopez, Daniel Ma, William Majoros, Joe McDaniel, Sean Murphy, Matthew Newman, Trung Nguyen, Ngoc Nguyen, Marc Nodell, Sue Pan, Jim Peck, Marshall Peterson, William Rowe, Robert Sanders, John Scott, Michael Simpson, Thomas Smith, Arlan Sprague, Timothy Stockwell, Russell Turner, Eli Venter, Mei Wang, Meiyuan Wen, David Wu, Mitchell Wu, Ashley Xia, Ali Zandieh, and Xiaohong Zhu. 2001. Sequencing the human genome. *Science*, 291(5507): 1304–1351. DOI: 10.1126/science.1058040. 84

[VeDC17]   J. S. Vetter, E. P. DeBenedictis, and T. M. Conte. 2017. Architectures for the post-Moore era. *IEEE Micro*, 37(4): 6–8. DOI: 10.1109/MM.2017.3211127. 118

162 **REFERENCES**

[VoSa11]   Panagiotis D. Vouzis and Nikolaos V. Sahinidis. 2011. GPU-BLAST: using graphics processors to accelerate protein sequence alignment. *Bioinformatics*. Oxford Academic Journal, Oxford, UK, 27(2): 182–8. DOI: 10.1093/bioinformatics/btq644. 85

[VSBG18]   Sudharshan S. Vazhkudai, Bronis R. de Supinski, Arthur S. Bland, Al Geist, James Sexton, Jim Kahle, Christopher J. Zimmer, Scott Atchley, Sarp Oral, Don E. Maxwell, Veronica G. Vergara Larrea, Adam Bertsch, Robin Goldstone, Wayne Joubert, Chris Chambreau, David Appelhans, Robert Blackmore, Ben Casses, George Chochia, Gene Davison, Matthew A. Ezell, Tom Gooding, Elsa Gonsiorowski, Leopold Grinberg, Bill Hanson, Bill Hartner, Ian Karlin, Matthew L. Leininger, Dustin Leverman, Chris Marroquin, Adam Moody, Martin Ohmacht, Ramesh Pankajakshan, Fernando Pizzano, James H. Rogers, Bryan Rosenburg, Drew Schmidt, Mallikarjun Shankar, Feiyi Wang, Py Watson, Bob Walkup, Lance D. Weems, and Junqi Yin. 2018. The design, deployment, and evaluation of the CORAL pre-exascale systems. In *Proceedings of the International Conference for High Performance Computing, Networking, Storage, and Analysis (SC '18)*. IEEE Press, Article 52, 1–12. 104, 105[Wass15] Shawn Wassermen. 2015. CAE Industry Experts Predict Future of Simulation. Retrieved Dec. 18, 2019 from https://www.engineering.com/DesignSoftware/DesignSoftwareArticles/ArticleID/10434/CAE-Industry-Experts-Predict-Future-of-Simulation.aspx. 81, 120

[Wats72]   W. J. Watson. 1972. The TI ASC: a highly modular and flexible super computer architecture. In *Proceedings of the December 5-7, 1972, Fall Joint Computer Conference, Part I (AFIPS '72 (Fall, part I))*. ACM, New York, 221–228. DOI: 10.1145/1479992.1480022. 17

[WHwu11]   Wen-mei Hwu. (Ed.) 2011. *GPU Computing Gems Emerald Edition*. Morgan Kaufmann. 48

[Widd80]   L.C. Widdoes. 1980. *The S-1 Project: Developing High-Performance Digital Computers*. IEEE Computer Society: COMPCOM Spring. 16

[WikCray]   Wikipedia. Seymour Cray. Retrieved Aug. 2020 from https://en.wikipedia.org/wiki/Seymour_Cray. 11

[WikCTS]   Wikipedia. Cray time sharing system. June 4, 2019. Retrieved Dec. 18, 2019 from https://en.wikipedia.org/wiki/Cray_Time_Sharing_System. 59

[WikTCA]   Wikipedia. Timeline of computer animation in film and television. Retrieved Dec. 18, 2019 from https://en.wikipedia.org/wiki/Timeline_of_computer_animation_in_film_and_television. 83

[WikNAG]   Wikipedia. NAG numerical library. Retrieved Dec. 18, 2019 from https://en.wikipedia.org/wiki/NAG_Numerical_Library. 66

[WikLCM]   Wikipedia. List of computer museums. Retrieved Dec. 18, 2019 from https://en.wikipedia.org/wiki/List_of_computer_museums. 134

[WikSUM]   Wikipedia. Summit (supercomputer). Retrieved Dec. 18, 2019 from https://en.wikipedia.org/wiki/Summit_(supercomputer). 134

[WikWHI]   Wikipedia. Whirlwind I. Retrieved Dec. 18, 2019 from https://en.wikipedia.org/wiki/Whirlwind_I. 134

[Wils94]   Gregory V. Wilson. 1994. *The History of the Development of Parallel Computing*. Retrieved Dec. 14, 2019 from https://webdocs.cs.ualberta.ca/~paullu/C681/parallel.timeline.html. 134

[WNNC17]   Jian-Ping Wang, Michael Niemier, Azad Naeemi, Chia-Ling Chien, Caroline Ross, Roland Kawakami, Sachin Sapatnekar, C.H. Kim, P. Crowell, Steve Koester, Supriyo Datta, Kaushik Roy, Anand Raghunathan, and X. Hu. (2017). A pathway to enable exponential scaling for the beyond-CMOS era: Invited. 1-6. DOI: 10.1145/3061639.3072942. 117

[Wolf15]   Michael Wolfe. 2015. Compilers and more: Is Amdahl's Law still relevant? *HPCWire* (January 2015). Retrieved Dec. 9, 2018 from https://www.hpcwire.com/2015/01/22/compilers-amdahls-law-still-relevant/. 129

[WSSA18]   Gulshan, Wadhwa, P. Shanmughavel, Singh, Atul Kumar, and Jayesh Bellare (Eds.). 2018. *Current Trends in Bioinformatics: An Insight*. Springer, Singapore, Singapore. 85

[YBAA20]   K. Yelick, A. Buluç, M. Awan, A. Azad, B. Brock, R. Egan, S. Ekanayake, M. Ellis, E. Georganas, G. Guidi, S. Hofmeyr, O. Selvitopi, C. Teodoropol, and L. Oliker. (2020). The parallelism motifs of genomic data analysis. *Philosophical Transactions of the Royal Society A: Mathematical, Physical, and Engineering Sciences*, 378(2166): 1–17. DOI: 10.1098/rsta.2019.0394. 85

[YCYG09]   Kisun You, Jike Chong, Youngmin Yi, Ekaterina Gonina, Christopher J. Hughes, Yen-Kuang Chen, Wonyong Sung, and Kurt Keutzer. 2009. Parallel scalability in speech recognition. *IEEE Signal Processing Magazine*, 26(1): 124–135. DOI: 10.1109/MSP.2009.934124. 92

[YuMa17]   Wenbo Yu and Alexander D. MacKerell. 2017. Computer-aided drug design methods. In P. Sass (eds) *Antibiotics. Methods in Molecular Biology*, vol 1520. Humana Press, New York, 85–106. DOI: 10.1007/978-1-4939-6634-9_5. 88

[YSHK85]   Toshitsugu Yuba, Toshio Shimada, Kei Hiraki, and Hiroshi Kashiwagi. 1985. SIGMA1: A dataflow computer for scientific computations. *Computer Physics Communications*, 37(1–3): 141–148. DOI: 10.1016/0010-4655(85)90146-8. 23

[ZCZX08]   Q. Zhang, Y. Chen, Y. Zhang, and Y. Xu. 2008. Sift implementation and optimization for multi-core systems. *Proceedings of the IEEE International Symposium on Parallel and Distributed Processing (IPDPS'08)*, 1–8. DOI: 10.1109/IPDPS.2008.4536131. 89

[Zhan12]   Xiangyang Zhang. 2012. The application of GPU system in the field of seismic data processing. *HPC Advisory Council Switzerland Conference 2012*, March 2012. Retrieved Dec. 14, 2019 from http://www.hpcadvisorycouncil.com/events/2012/Switzerland-Workshop/Presentations/Day_3/8_CNPC.pdf. 78

[ZLZM11]   Yan Zhai, Mingliang Liu, Jidong Zhai, Xiaosong Ma, and Wenguang Chen. 2011. Cloud versus in-house cluster: evaluating Amazon cluster compute instances for running MPI applications. In *State of the Practice Reports (SC '11)*. ACM, New York, Article 11, 10 pages. DOI: 10.1145/2063348.2063363. 84

[Zwak75]   R. G. Zwakenberg. 1975. Vector extensions to LRLTRAN. Proceedings of the conference on Programming languages and compilers for parallel and vector machines. ACM, New York, 77-86. DOI: 10.1145/800026.80840. 57

Authors' Biographies

Robert Kuhn received his Ph.D. from the University of Illinois at Urbana-Champaign in 1981. In 1983, as assistant professor at Northwestern University, he consulted on the vector register architecture for the Gould SEL real-time minicomputers. In 1987, he led Alliant Computer System's vectorizing-parallelizing compiler team. In 1990, he led the team of application experts at Alliant. In 1992 when Alliant closed, he worked for Kuck and Associates, Inc. and led the customer experts where, for example, he worked with SGI and other OEMs on the definition and adoption of OpenMP. In 2000 when Intel acquired KAI, he worked on adoption and integration of threading by HPC ISVs He managed the acquisition by Intel of Pallas GmbH and their MPI tools. He managed Intel's participation in the ASCI/LLNL Ultrascale project to develop MPI/OpenMP performance analysis tools and led development of other Intel HPC tools. Dr. Kuhn led the adoption of threading by ISVs for the introduction of Intel's first multicore processor and the Intel/Microsoft Universal Parallel Computing Research Center project with University of California Berkeley and University of Illinois at Urbana-Champaign, as well as managing approximately 20 other university research projects in high performance computing.

David Padua received his Ph.D. from the University of Illinois at Urbana-Champaign in 1980. In 1985, after a few years at the Universidad Simón Bolívar in Venezuela, he returned to the University of Illinois where he is now Donald Biggar Willet Professor in Engineering. He has served as program committee member, program chair, or general chair to more than 70 conferences and workshops. He was the Editor-in-Chief of Springer-Verlag's *Encyclopedia of Parallel Computing* and is currently a member of the editorial board of the *Communications of the ACM*, the *Journal of Parallel and Distributed Computing*, and the *International Journal of Parallel Programming*.

Dr. Padua has supervised the dissertations of 30 Ph.D. students. He has devoted much of his career to the study of languages, tools, and compilers for parallel computing and has authored or co-authored more than 170 papers in these areas. He received the 2015 IEEE Computer Society Harry H. Goode Award. In 2017, he awarded an honorary doctorate from the University of Valladolid in Spain. He is a Fellow of the ACM and the IEEE.

Printed in the United States
by Baker & Taylor Publisher Services